Mathematics for Health Occupations

Kathi A. Dunlap

Delmar Publishers Inc.

NOTICE TO THE READER

Delmar staff:
Senior Executive Editor: David Gordon
Project Supervisor: Susan L. Simpfenderfer
Design Supervisor: Susan Mathews
Art Coordinator: Michael Nelson

For more information, address Delmar Publishers Inc.
3 Columbia Circle, Box 15-015
Albany, New York 12212-5015

Printed in the United States of America
Published simultaneously in Canada
by Nelson Canada,
a division of The Thomson Corporation

10   9   8   7   6   5   4   3

Library of Congress Cataloging-in-Publication Data
Dunlap, Kathi A.
    Mathematics for health occupations / Kathi A. Dunlap.
      p. cm.
    Includes index.
    ISBN 0–8273–4173–3
    1. Arithmetic.   I. Title.
    [DNLM: 1. Health Occupations. 2. Mathematics.]
QA107.D86 1991
513—dc20
DNLM/DLC
for Library of Congress   90–3896
                           CIP

Preface

Mathematics for Health Occupations was written to meet a variety of needs in the field of health occupations. It is highly recommended for use in high school vocational programs, two-year nursing programs at a technical school or school of nursing, or in-house training programs offered by hospitals or nursing homes.

It can be used in a high school vocational course as a full-year mathematics program or on the college level as a one-quarter or one-semester mathematics course. I would also recommend integrating this into a health skills class where mathematics is not normally taught as a separate discipline and covering the mathematics material as it coincides with the health instruction. This book would also work well in a laboratory class situation where hands-on instruction is emphasized and workbook material is used to supplement the learning experience.

Organization

The book begins with an emphasis on remedial or review math skills. Students who are lacking in these basic skills will have great difficulty with accuracy and understanding in other chapters of the book. These beginning chapters (1–6) may be used in class as a foundation for other chapters, as a review at various points throughout the book, or as individualized material for students who need remediation.

The next six chapters (7–12) are specialized to teach specific skills needed in health occupations. These chapters cover measurement, the metric system, roman numerals, medication dosages, vital signs, and intake and output. The text covers the necessary math skills needed to master each area.

The measurement (standard and metric) chapters are presented in a clear, easy-to-understand format that uses understanding, rather than simple memorization, as its basis. They include information on height, weight, and volume measurement—all of which are important in the health occupations.

Roman numerals are presented as a set of simple rules to follow, then much practice is given using the rules. This chapter was included to familiarize students with roman numerals so they can read prescriptions and physician's orders. Roman numerals are also useful in note taking at the high school or college level.

Chapter 10, "Medication Dosages," is full of information as well as instruction on computing dosages. It includes the basic rules for adult and children's dosages and offers a logical, systematic method for solving these problems. Again, emphasis is placed on understanding the use of the formulas (through ratios) rather than on memorization.

The chapter on vital signs (chapter 11) covers temperature, pulse, respiration, and blood pressure. It includes reading the various medical instruments (thermometer, sphygmomanometer, etc.) as well as charting the results. A section on temperature conversion is also included.

Chapter 12 is an in-depth look at intake and output. This chapter explains the various types of intake and output that are recorded and how to fill out an intake and output chart.

The last section of the book (chapters 13–17) covers an area that is too often overlooked—money management for both the office and the individual and office skills. Many students who are trained in health occupations will be required to deal with some aspect of office management, whether it is filing patient information or collecting money from a patient. These chapters address this need and others pertinent to office and personal management.

Money is the focus of chapter 13. In this chapter, the student will learn how to handle money correctly, how to make change, how to collect money from patients, and how to balance the cash drawer at the end of the day.

Chapter 14 covers time sheets and elapsed time. This is a necessary skill for determining the hours worked in a week on the job.

The next chapter (15) goes further with this concept and helps the student determine his gross and net pay based on hours worked. Salary and commission are also covered as types of earnings. Once net pay is computed, the student then has the opportunity to write paychecks. This makes the skills learned in chapters 14 and 15 beneficial not only for writing individual paychecks but also for the student who may be responsible for computing payroll on the job.

Chapter 16 covers other office skills that a student would use in an office or hospital setting. Numerical filing, appointment scheduling, and receipt writing are explained and practiced in depth.

The last chapter offers the student a clear understanding of personal finance. This chapter includes checking and savings accounts, the use of credit, utility bills, purchasing a car, renting an apartment, and establishing a monthly budget.

Acknowledgments

I would like to thank the reviewers of this book for their comments, suggestions, and encouraging remarks. These reviewers include:

Harold Green
Orangeburg Calhoun Tech
3250 St. Mathews Road N.E.
Orangeburg, SC 29115

Allene Shelton
Cerro Casa Community College
3000 College Heights Blvd.
Ridgecrest, CA 93555

Susan Stokley
Spartanburg Technical College
I85 and New Cut Road
Spartanburg, SC 29305

Sharon Vansant
California Paramedical and Technical College
4550 La Sierra
Riverside, CA 92505

I also would like to thank David Gordon for his positive encouragement and suggestions. His help has made this dream become a reality.

To my loving husband, Vern, and daughter, Danielle. You've sacrificed much to see me through this project. Thank you for your support and your love. PTL!

Contents

Chapter 1

Place Values

The place values in our number system are:

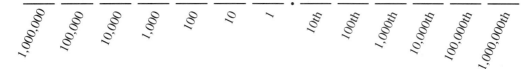

Knowing these place values is important in many mathematical operations such as rounding numbers, comparing numbers, reading numbers, or lining up columns to add. It is also important when writing out numbers as words for checks or receipts.

The places to the right of the decimal all end in "th" and are **part** of a whole number. The places to the left of the decimal all indicate multiples of a whole number. When a number does not have a decimal in it, assume that the decimal is at the right. For example, 74 is understood to mean 7 tens and 4 ones, and the decimal belongs to the right (74.).

Problems State the place value of the underlined number.

1) 724.<u>8</u> tenth	11) 46.898<u>9</u> _____
2) 1<u>0</u>,429 thousand	12) 4<u>6</u>,892 _____
3) 4.3<u>6</u>71 hundredth	13) 8.<u>0</u>01 _____
4) <u>9</u>.83 one	14) 47.7<u>9</u>6 _____
5) <u>2</u>14.53 hundreds	15) <u>5</u>21 _____
6) <u>1</u>00,000.1 thousands	16) 842<u>6</u>2 _____
7) <u>3</u>6495 ten thousands	17) <u>3</u>9.417 _____
8) 8<u>6</u> ones	18) 8,<u>6</u>13,218.4 _____
9) 4.6<u>7</u>5 hundredths	19) 1,3<u>7</u>6,582 _____
10) 82.00<u>1</u> thousandths	20) 342<u>9</u>.63 _____

Indicate which of the digits (0–9) are in the specified place value in this number: 9,876,543.210

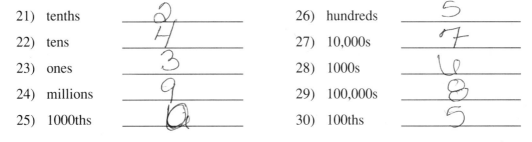

21) tenths	2	26) hundreds	5
22) tens	4	27) 10,000s	7
23) ones	3	28) 1000s	6
24) millions	9	29) 100,000s	8
25) 1000ths	0	30) 100ths	5

Comparing Numbers

Comparing numbers is necessary when making decisions about a patient's care. It is also useful in many office skills, such as numerical filing, ordering supplies, figuring parts of a payroll, and so on.

To compare numbers, first line up the decimals. This is a process of putting certain place values of one number above the same place value of the other number or numbers.

117.68   The ones (for example) are on top of each other.
111.452

Then, beginning from left to right, compare like place values. The digits in the hundreds place in both numbers are the same. The digits in the tens place in both numbers are the same. The ones place is where the first difference occurs. Since the 1 in 11**1**.452 is smaller than the 7 in 11**7**.68, 111.452 is the smaller number.

Sometimes the numbers you are comparing do not have the same number of place values. When this occurs, the number that holds the highest place value is the largest. For example,

11768.2   holds the ten thousands place
1176.82   holds the thousands place

Therefore, 11768.2 is larger than 1176.82, because the ten thousands place is larger than the thousands place.

Example 1

Which is larger?   0.**7**0
0.713

Looking left to right, the first digits that are different are in the hundredths place. Since 1 is larger than nothing at all, 0.713 is larger.

Example 2

Which is larger?   2960.5
2956.83

Figure 1–1   It is necessary to compare numbers when filing.

Looking left to right, the first digits that are different are in the tens place. Since 6 is larger than 5, 2960.5 is larger than 2956.83.

Example 3

Which is larger?   .1
                   .0602

Looking left to right, the 0 and 1 in the tenths place are different. Since 1 is larger than 0, .1 is larger than .0602.

Problems   Line up the place values before comparing numbers. Circle the larger of the two numbers.

1) 4.786	11) 3	21) 151.1
4.7	15	111
2) 5.39	12) 89	22) 36.8
53	90	36.799
3) 6.002	13) 2173	23) 2577
6.2	220.3	527
4) 40,000	14) 5.001	24) 38.96
4962	5.0	39.01
5) 73.6	15) 6159	25) 2.65
700	6200	2.069
6) 9.209	16) 8.4	26) 51
9.21	9.1	50.8
7) 468	17) 26.3	27) 776
486	.795	7676
8) 6.3	18) 100	28) 98
6.03	100.06	907.9
9) 49288	19) 53	29) 8010
49300	52.869	8100
10) 8.21	20) 416	30) 453.2
8.4	42.9	2545

Circle the number that is not equal to the others.

31) 8.010	32) 53.6	33) 1000.2	34) 57.606	35) .001
8.100	53.60	1000.20	57.6060	.0010
8.01	53.06	1002.0	57.60600	.1010
8.01000		1000.200	57.60660	0.001

Rounding Numbers

To round numbers to a certain place value, it is first necessary to determine where that place value is. Then you can use the number directly to the right of that place value to decide whether or not to round the original place value. If the right digit is 5 or larger, round up the original digit. If it is less than 5, leave the original digit alone.

Example 1

Round 5.9267 to the 100th place.

1) Where is the 100th place?   5.9267

2) Use the digit to the right to decide.   5.9267

3) Since 6 is in the "5 or larger" category, round the 2 up. (Drop off any remaining digits or change them to zeros.)

4) Solution is: 5.93

Example 2

Round 531.89 to the 100s place.

1) Where is the 100s place?   531.89

2) Use the digit to the right to decide.   531.89

3) Since 3 is in the "less than 5" category, leave the 5 alone. Change remaining digits to zeros.

4) Solution is: 500.00 or 500. (Since these are equal numbers, either solution is correct.)

Example 3

Round 57.38 to the nearest unit (or ones) place.

1) Where is the ones place?   57.38

2) Use the digit to the right to decide.   57.38

3) Round up or leave alone? Leave alone.

4) Solution is: 57.00 or 57

Problems   Round the following numbers to the specified place value.

Round to the nearest 1000.

1) 57,263	_____	4) 3,965	_____	7) 876,521	_____
2) 981	_____	5) 2175	_____	8) 430,020	_____
3) 3176.5	_____	6) 4616.99	_____	9) 27,647,531	_____

Round to the nearest 100.

1) 57,263	_____	4) 3,965	_____	7) 876,521	_____
2) 981	_____	5) 2175	_____	8) 430,020	_____
3) 3176.5	_____	6) 4616.99	_____	9) 27,647,531	_____

Round to the nearest 10.

1) 57,263	_____	4) 3,965	_____	7) 876,521	_____
2) 981	_____	5) 2175	_____	8) 430,020	_____
3) 3176.5	_____	6) 4616.99	_____	9) 27,647,531	_____

Round to the nearest 1.

1) 4.77653	_____	4) 81.4753	_____	7) 4.00605	_____
2) 7.4939	_____	5) 213.6168	_____	8) 2.7893	_____
3) 51.673	_____	6) 11.1111	_____	9) 5.1515	_____

Round to the nearest 10th.

1) 4.77653	_____	4) 81.4753	_____	7) 4.00605	_____
2) 7.4939	_____	5) 213.6168	_____	8) 2.7893	_____
3) 51.673	_____	6) 11.1111	_____	9) 5.1515	_____

Round to the nearest 100th.

1) 4.77653	_____	4) 81.4753	_____	7) 4.00605	_____
2) 7.4939	_____	5) 213.6168	_____	8) 2.7893	_____
3) 51.673	_____	6) 11.1111	_____	9) 5.1515	_____

Round to the nearest 1000th.

1) 4.77653	_____	4) 81.4753	_____	7) 4.00605	_____
2) 7.4939	_____	5) 213.6168	_____	8) 2.7893	_____
3) 51.673	_____	6) 11.1111	_____	9) 5.1515	_____

Round to the appropriate place value.

10th	1) 4.796 _____	2) 34.87 _____	3) 2.115 _____
ones	4) 89.65 _____	5) 4.332 _____	6) 8.5 _____
hundreds	7) 2119.4 _____	8) 806.5 _____	9) 21,768 _____
1000ths	10) 4.11654 _____	11) 3.7547 _____	12) .00129 _____
tens	13) 147.616 _____	14) 81.4 _____	15) 363.83 _____
100ths	16) 5.776 _____	17) 8.9999 _____	18) 236.417 _____

Reading and Writing Numbers as Words

To read numbers out loud correctly, it is helpful to separate the number into parts. It is also necessary to know the place values learned in the previous section.

Example 1

Read 6027498.1.

1) First, separate the left side into the traditional units by commas (every three digits from the decimal). 6027498.1 becomes 6,027,498.1

2) Next, read each section separately, stating the digits and the place value of the last digit. Hundreds are usually stated whenever numbers are in groups of three, as seen in this example.

	Say...
<u>6</u>,	6 million
02<u>7</u>,	27 thousand
49<u>8</u>	498 (omit saying "ones" and say "and" instead)
.1	1 tenth

or 6 million, 27 thousand, 4 hundred 98 and 1 tenth.

Thus, the word "and" separates the whole number (left side) from the part of a whole number (right side). Do not say "and" if there is nothing on the right side.

Example 2

Read 4965.821.

1) 4965.821 becomes 4,965.821

2) Next

	Say...
<u>4</u>,	4 thousand
965	965 "and"
.821	821 thousandths

or 4 thousand, 9 hundred 65 and 8 hundred 21 thousandths.

Example 3

Read .004678.

1) Nothing to separate with commas

2) Next

	Say...
.004678	4,678 millionths

or 4 thousand, 6 hundred 78 millionths.

Example 4

Read 25689.004201.

1) 25689.004201 becomes 25,689.004201

2) Next

	Say...
2<u>5</u>,	25 thousand
689	689 "and"
.004201	4,201 millionths

or 25 thousand, 6 hundred 89, and 4 thousand 2 hundred one millionths.

Problems   Read the following numbers orally.

1)	4296.81	6)	475.31	11)	53274.83
2)	216593	7)	81267.5832	12)	4.76
3)	74.659	8)	772.568	13)	1.005
4)	9.8	9)	27.0003	14)	.45
5)	3	10)	270.03	15)	.0045

Write these numbers as shown in the examples.

16) 462.373

17) 17344.47

18) 274.482

19) 1694869

20) .000306

21) 37.97

22) 163.66

23) 882.22

24) 7390

25) 23.6375

26) 18.337

27) 28587

28) 1563.333

29) 16.33685

30) 2685.9

Chapter 2

Whole Number Operations

Mathematics is an important skill in the health occupations. Accuracy can mean the difference between life and death in computing dosages of medicine, recording vital signs, and other health areas. Being competent in addition, subtraction, multiplication, and division is the basis for accuracy in all other mathematical operations.

Addition

To add whole numbers, it is important to line up the numbers according to place values. Neatness in lining up the columns will ensure an accurate total.

$$
\begin{array}{r}
2 \\
108 \\
13 \\
17 \\
22 \\
+\ 7 \\
\hline
167
\end{array}
$$

- Add the ones column.
  $8 + 3 + 7 + 2 + 7 = 27$
- Write down the 7 and carry the 2 over to the tens column.
- Add the tens column.
  $2 + 1 + 1 + 2 = 6$
- Write down the 6 and carry nothing.
- Add the hundreds column, write down, carry if necessary.
- Continue until all columns are totaled.

Figure 2–1 Reprinted with permission from Rice, *Medications and Mathematics for the Nurse*, 6E, copyright 1988, Delmar Publishers Inc., Albany, NY.

Your addition can be greatly simplified by grouping numbers as you add. Let's look at the ones column again.

108	$8 + 2 = 10$	• Add any numbers together that together equal ten.
13	$3 + 7 = 10$	• Then add all remaining numbers in. (solution: 27)
17	$+ 7$	
22	$\overline{27}$	
$+ 7$		
$\overline{167}$		

Numbers that add together to equal 10 are called *complementary*. These include:

$1 + 9$
$2 + 8$
$3 + 7$
$4 + 6$
$5 + 5$

When you are adding, look for pairs like this to simplify your work.

Problems Solve the following problems by lining up the numbers to be added in a vertical column and then completing the addition.

1) $2 + 6 + 5 + 9 + 1 + 8 =$ _____

2) $1 + 3 + 4 + 7 + 9 + 6 + 4 =$ _____

3) $8 + 3 + 9 + 7 + 2 =$ _____

4) $6 + 3 + 2 + 8 + 4 =$ _____

5) $14 + 26 + 3 =$ _____

6) $18 + 19 + 22 + 6 =$ _____

7) $6 + 15 + 34 + 8 + 25 =$ _____

8) $4 + 7 + 18 + 6 + 33 =$ _____

9) $3 + 9 + 11 + 27 + 4 =$ _____

10) $8 + 6 + 9 + 3 + 4 + 7 =$ _____

11)
```
  4678
+  481
_____
```

12)
```
 36,158
+ 2,488
_____
```

13)
```
 4,846
+  333
_____
```

14)
```
  864
  821
  916
+ 799
_____
```

15)
```
   413
   216
   744
+  316
_____
```

16)
```
   93
   81
   67
+  74
_____
```

17)
```
   81
   12
   17
+  93
_____
```

18)
```
  3,461
    829
    913
+   875
_____
```

19)
```
  63,996
  17,963
+  8,414
_____
```

20)
```
  189,641
+ 328,479
_____
```

21) Janice is required to take inventory in Dr. Madison's office every month. She counts the supplies in each examination room and adds them to find a total inventory. Total the following:

tongue depressors:	127	boxes of cotton	8
	63		3
	219		2
	151		4
	18		11

examination gloves	586	thermometer covers	381
	211		218
	394		157
	88		108
	115		93

22) Janice also has to take inventory of the reception desk and filing area. Total the following:

file folders:	red	1,681
	blue	468
	green	852
	yellow	1,005

labels	69
	1,488
	206
	311
	2,044
	817

forms:	prescriptions	3,606
	doctor's excuses	981
	charts	2,804
	medical history	1,963

Figure 2–2

23) Craig works at Parkview Care Center as a nurse's aide. It is his responsibility to total the intake and output records of several patients. Total the following intake and output amounts (all answers are in cc—cubic centimeters).

Intake:

Oral	IV	Irrigation
550	850	20
380	965	15
1,110	1,080	35
480		60
120		
120		

Total Intake: _____

Output:

BM	Emesis	Urine	Irrigation
210	160	250	50
350	500	460	85
		185	35
		490	

Total Output: _____

24) Next patient's intake and output is as follows:

Intake:

Oral	IV	Irrigation
460	805	15
280	300	10
100	260	25
30		40
820		10

Total Intake: _____

Output:

BM	Emesis	Urine	Irrigation
300	550	360	50
250	110	425	65
400		405	
		510	

Total Output: _____

25) BONUS: Find the digits that A and B represent.

    1,A65,328   A = _____

    + 476,83A   B = _____

    B,44B,167

Subtraction

To subtract whole numbers, it is important to line up the numbers according to place values. Neatness in lining up the columns is essential.

$$\begin{array}{r} {}^{14} \\ 0\cancel{1}7 \\ -\ 63 \\ \hline 84 \end{array}$$

- Subtract the ones column, if possible.
  $$7 - 3 = 4$$
- Subtract the tens column, if possible.
  $$4 - 6 = \text{impossible}$$
- When subtraction cannot be done, borrow 1 from the next column and use it in your subtraction.
  $$14 - 6 = 8$$

Sometimes you must borrow from more than one column. Let's look at this example:

$$\begin{array}{r} {}^{3\ 9\ 10} \\ \cancel{400} \\ -\ 97 \\ \hline 303 \end{array}$$

- Subtract the ones column, if possible.
  $$0 - 7 = \text{impossible}$$
- When subtraction cannot be done, borrow 1 from the next column and use it in your subtraction. But in this case, 1 can't be borrowed from the 0, so you must borrow from the third column.
- Borrowing 1 from 4 in the hundreds column leaves a 3 and makes the 0 in the tens column become 10.
- Borrowing 1 from that 10 leaves a 9 (shown) and makes the 0 in the ones column become a 10.
- Subtraction may now be completed:

  ones $10 - 7 = 3$

  tens $9 - 9 = 0$

  hundreds $3 - 0 = 3$

Problems   Line up all problems vertically before subtracting.

1) $14 - 8 =$ _____

2) $27 - 19 =$ _____

3) $109 - 46 =$ _____

4) $67 - 23 =$ _____

5) $89 - 21 =$ _____

6) $115 - 47 =$ _____

7) $41 - 9 =$ _____

8) $53 - 36 =$ _____

9) $101 - 74 =$ _____

10) $18 - 6 =$ _____

11) $\begin{array}{r} 721 \\ -\ 46 \\ \hline \end{array}$

12) $\begin{array}{r} 48,610 \\ -\ 396 \\ \hline \end{array}$

13) $\begin{array}{r} 3,917 \\ -\ 86 \\ \hline \end{array}$

14) $\begin{array}{r} 1,964 \\ -\ 128 \\ \hline \end{array}$

15)   468
        − 27

17)   682
        − 96

19)   409
        − 89

16)   3,939
      − 984

18)   8,416
      − 767

20)   1,047
      − 953

21) a. Janice uses the inventory she takes every month to figure out how many supplies have been used. If she started with 1,683 tongue depressors and is left with 578, how many were used?   _____

     b. If she started with 47 boxes of cotton and is left with 28, how many were used?   _____

     c. If she started with 2,116 examination gloves and is left with 1,394, how many were used?   _____

     d. If she started with 1,463 thermometer covers and is left with 957, how many were used?   _____

22) a. Janice is required to keep a certain amount of all reception area supplies in stock. If she must keep 5,000 file folders on hand and her inventory shows 4,006, how many more should she order?   _____

     b. If she must keep 5,000 labels on hand and her inventory shows 4,935, how many more should she order?   _____

     c. If she must keep 5,000 of each of the following forms on hand, how many more of each should she order:

Prescriptions	3,606	_____
Doctor's Excuse	981	_____
Charts	2,804	_____
Medical History	1,963	_____

Figure 2–3

23) Jeff is employed at Bassill Family Whole Health Center full-time. He is paid regular wages for a 40-hour week and overtime for all hours worked over 40. How many hours overtime did he work each week?

       Week 1:   51 hours   _____

       Week 2:   43 hours   _____

       Week 3:   47 hours   _____

       Week 4:   56 hours   _____

24) Sharon keeps a record of how many infants are admitted and discharged from the nursery. On Monday, there were 3 infants in the nursery. On Tuesday, 4 were discharged, and on Wednesday, 7 more were released. How many infants remain in the nursery?   _____

25) BONUS: Find the digit that A represents.

      6417   A = _____

   – 3A8
   ―――――
    601A

Multiplication

To multiply whole numbers, it is important to line up the numbers according to place values. Neatness in lining up the columns is essential.

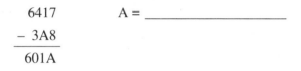

     146
    × 29
  ――――
  1314
   292
 ――――
 4234

- Multiply 9 times each digit in the top number.

    $9 \times 6 = 54$ Write 4, carry 5.

    $9 \times 4 = 36 + 5 = 41$ Write 1, carry 4.

    $9 \times 1 = 9 + 4 = 13$ Write 3, carry 1.

                    Write 1 carried.

- Multiply 2 times each digit in the top number. Begin writing under tens column.

    $2 \times 6 = 12$ Write 2, carry 1.

    $2 \times 4 = 8 + 1 = 9$ Write 9.

    $2 \times 1 = 2$ Write 2.

- Add the products.

Problems   Line problems up vertically before multiplying.

1) $7 \times 10 =$ _____

2) $13 \times 9 =$ _____

3) $11 \times 6 =$ _____

4) $5 \times 7 =$ _____

5) $14 \times 3 =$ _____

6) $19 \times 7 =$ _____

7) $15 \times 15 =$ _____

8) $12 \times 8 =$ _____

9) $16 \times 12 =$ _____

10) $20 \times 14 =$ _____

11)   176
     × 92
    ――――

12)   1,487
   × 963
  ――――

13)   3,572
 × 1,681
 ――――

14)   293
      × 47

15)   864
      × 777

16)   2,921
      × 2,786

17)   147
      × 39

18)   65,814
      × 21,493

19)   4,000
      × 2,867

20)   4,378
      × 2,913

21) Steve records all medications taken by his patients. If patients take an equal dosage several times a day, the dosage can be multiplied by the frequency to determine total medication. Find the amount of medication administered to the following patients.

   a. Bill Houser   40 mg   8 times a day   _____
   b. Rachel Brown   20 mg   4 times a day   _____
   c. Stephanie All   35 mg   6 times a day   _____
   d. Agnes Schultz   47 mg   5 times a day   _____
   e. Dave Woods   55 mg   24 times a day   _____

22) Shelly Hayes works in a hospital billing office. The room rates are posted on a sign in the billing office (see figure 2–4). Find the amount that should be billed to these patients for their rooms:

   a. Dave Wray   regular room   2 days   _____
   b. Penny Mikells   regular room   4 days   _____
   c. Justin Landers   ICU room   3 days   _____
   d. Kathy Krandall   regular room   10 days   _____
   e. Jack Mullin   ICU room   7 days   _____

23) This month, Helen is scheduled to work 14 8-hour shifts and 7 12-hour shifts. How many hours will she work this month?   _____

24) Dr. Graham bought new wall decorations for her examination rooms. She bought 6 pictures that cost $21 each, 4 shelves that cost $12 each, and 3 poster prints at $7 each. How much did she spend?   _____

Room Rates

Regular	$212 per day
ICU	$689 per day

Figure 2–4

Division

To divide whole numbers, place the dividend (number to be divided) under the division sign. Place the divisor (number you're dividing by) to the left of the sign.

Example 1

$$\begin{array}{r} 27 \\ 64\overline{)1728} \\ -128 \\ \hline 448 \\ -448 \\ \hline 0 \end{array}$$

- Look at the dividend from left to right, choosing a number 64 will divide into.
  - Will 64 divide into 1? no
  - Will 64 divide into 17? no
  - Will 64 divide into 172? yes
- Decide how many times 64 will to into 172 (64 × 2 = 128).
- Place the 2 above the last digit in 172 (the 2).
- Write down the 128 and subtract.
- Bring down the next digit after 172 (the 8) and decide how many times 64 will go into 448 (64 × 7 = 448).
- Place the 7 above the 8 in 1728.
- Write down the 448 and subtract.
- In whole number division, the difference after all numbers are brought down is called the *remainder*.
- In this case, the remainder is 0 and is not written.

Example 2

$$\begin{array}{r} 541 \;\; R\,21 \\ 27\overline{)14628} \\ -135 \\ \hline 112 \\ -108 \\ \hline 48 \\ -27 \\ \hline 21 \end{array}$$

- Look at the dividend from left to right, choosing a number 27 will divide into.
  - Will 27 divide into 1? no
  - Will 27 divide into 14? no
  - Will 27 divide into 146? yes
- Decide how many times 27 will go into 146 (27 × 5 = 135).
- Place the 5 above the last digit in 146 (the 6).
- Write down the 135 and subtract.
- Bring down the next digit (2) and decide how many times 27 will go into 112 (27 × 4 = 108).
- Place the 4 above the 2 in 14628.
- Write down the 108 and subtract.
- Bring down the next digit (8) and decide how many times 27 will go into 48 (27 × 1 = 27).
- Place the 1 above the 8 in 14628.
- Write down the 27 and subtract. 21 is the remainder.

Problems   Complete the following division problems. Write the answer and any remainder.

1) 276 ÷ 40 _____   8) 389 ÷ 25 _____   15) 3608 ÷ 43 _____

2) 294 ÷ 14 _____   9) 956 ÷ 33 _____   16) 48960 ÷ 80 _____

3) 1346 ÷ 11 _____   10) 3296 ÷ 8 _____   17) 7225 ÷ 78 _____

4) 3336 ÷ 109 _____   11) 30098 ÷ 149 _____   18) 11568 ÷ 16 _____

5) 1250 ÷ 207 _____   12) 2910 ÷ 91 _____   19) 26460 ÷ 29 _____

6) 357 ÷ 17 _____   13) 415 ÷ 8 _____   20) 2550 ÷ 34 _____

7) 1388 ÷ 84 _____   14) 12289 ÷ 74 _____

21) Mrs. Pierson had her Pomeranian in a veterinary hospital for 6 days.
     The total room bill was $174. What was she charged for each day?   _____

22) John Shane works for a pharmaceutical company. He has 468 cases of
     acetaminophen to store in warehouse 2. If he can move 18 cases in his
     forklift at one time, how many trips will he have to make to move all
     of the cases?   _____

23) Sharon Heights is a dietician at Montpelier Hospital. As a research project,
     she is asked to determine the average daily sodium intake of the nurses on
     the cardiac ward at the hospital. To find the average, she adds all sodium
     intake levels together and divides by the number of nurses in the study.
     Find the average of the following:

2430	2110	2645	2500	2100
2165	2005	3105	2780	1980

     _____

24) Amanda Bierly read a physician's order to administer 1140 mg of
     medication over a 24-hour period. How much medication will she give
     at each dosage if she gives it every hour?   _____

Mixed Operations

Problems   Identify the operations needed to solve the following problems and then solve.

1) Janice works 8 hours a day, 5 days a week. She also works 5 hours on
    Saturday. How many hours does she work altogether?   _____

2) Brianne takes inventory at Arnold's Drug. In cold medicines, she has
    6 cases of Hoptifed, 3 cases of Nusil, and 5 cases of Chloral. Each case
    has 24 bottles in it.

    a. How many bottles of each does she have?   _____

    b. How many bottles does she have altogether?   _____

3) Rita works in housekeeping at Allen Creek Care Community. She has 5 aides under her supervision. If the aides have 165 rooms to clean, how many must each complete? _____

4) Jim needs to order new dining trays for the hospital. They need to have 2300 on hand, and they have 1863 at last count. How many should he order? _____

5) Teresa spends $182 on 13 pamphlet sets for the dental office. How much does each pamphlet set cost? _____

6) Alyse needs 130 credits to graduate from Mercy Hospital's School of Nursing. She currently has 119. How many more does she need? _____

7) Rick bought 7 new uniforms when he completed his EMT training. Each uniform cost $21. How much did he spend on uniforms? _____

8) Jessica Rayburn recorded intake and output for Mr. Carothers. In an 8-hour period, he drank 3 large glasses of water (240 cc each), 1 small bowl of broth (120 cc), and 1 glass of apple juice (100 cc).

   a. How much did his oral intake total? _____

   b. What was his average intake per hour? _____

9) Sherry missed 7 hours of work due to illness. If she normally works 51 hours a week, how many hours did she work this week? _____

10) Cathy needs 500 gowns in each of 8 examination rooms. She currently has 2938 in stock. How many more should she order? _____

Chapter 3

Decimal Operations

When working with decimal numbers in calculations, it is important to understand what significance the decimal has in the number.

In the number 211.65, two hundred eleven is a **whole** number. It represents 211 entire units (of money, if used with a dollar sign; of medication, if used with cc, etc.). Every whole number is merely a count of some sort of item. The .65 in 211.65 represents a **part** of a whole unit. For example, in money, $.65 means part of a dollar (indicated by the dollar sign). Sixty-five cents is part of a dollar, but not a whole dollar.

A common mistake is to write .65¢ to mean sixty-five cents. The decimal in .65 means a part of a whole, and the ¢ symbol means cents. Therefore, .65¢ means part of a cent, not 65 whole cents, but part of one cent. The correct way to write sixty-five cents is either $.65 (part of a dollar) or 65¢ (sixty-five whole cents).

Addition

To add decimal numbers, it is important to line up the numbers according to place value (commonly called "lining up the decimal"). Neatness in lining up the columns will help ensure an accurate total.

Once the numbers are lined up correctly, the process of addition is the same as in whole number addition.

Example 1

Add 21.061 + 4.08 + 161.3 + 49

	2	1.	0	6	1
		4.	0	8	
1	6	1.	3		
	4	9.			
2	3	5.	4	4	1

Notice that the decimal is placed in the answer in alignment with where it was in the individual numbers.

Problems   Write the following problems vertically before solving.

1) 8 + 2.06 + 5.353 + 9.99 + 100 =   _____

2) 41 + 8.602 + 9.009 + 4.16 =   _____

3) .0009 + .009 + .9001 + .009 + .081 = _____

4) 5 + .5 + .05 + .005 + .0005 = _____

5) 19 + 216.53 + 8.86 + .0051 = _____

6) 9.621 + 9001.001 + 46.259 + 8.0001 = _____

7) 1.03 + .8 + .052 + 9.96 + 211.4 = _____

8) 4.6 + 6.4 + 8.001 + .05 + 216.003 = _____

9) 19 + 99.99 + .009 + .001 = _____

10) 6 + 71.3 + 48.004 + .5 + .045 + .005 = _____

11) Find the total of the following bank deposit.

pennies	.29
nickels	2.05
dimes	1.10
quarters	8.75
$1 bills	48.00
$5 bills	20.00
$10 bills	60.00
$20 bills	80.00
checks	179.87

12) Janice Boop is a billing technician at Boston General Hospital. She totals pharmacy bills for patients being discharged. Find the total of the following pharmacy bill.

analgesics	$27.70
antibiotics	52.29

13) Complete the following invoice:

Quantity	Item	Unit Cost	Extension
3	case sterile gauze	$ 17.489	$ 52.467
6	box exam gloves	210.465	1,262.790
9	box tongue depressors	17.899	161.091
		Total =	

14) You work at Blyroy Hospital in inventory control. You need to total the following inventory for the Nurses' Station on 2–South.

desks	$ 679.98
secretarial chairs	308.81
wastebaskets	21.17
desk lamps	43.55
file cabinets	654.04
stapler	6.14
tape dispenser	3.16

15) Karen Harshbarger has an insurance rating of 3.97 on her car. This means she pays 3.97 times the base rate. Next year, she will buy a nicer car and her rating will go up by .0701. What is her new rating going to be? _____

Subtraction

In subtraction of decimal numbers, it is again important to line up the numbers according to place value.

Once the numbers are lined up correctly, the process of subtraction is the same as in whole number subtraction. One exception, however, is when the minuend (number from which you are subtracting) holds fewer place values than the subtrahend (number being subtracted).

Example 1

$$\begin{array}{r} 1.06 \\ -\,0.1742 \\ \hline \end{array}$$ Minuend has fewer place values.

$$\begin{array}{r} 1.06\mathbf{00} \\ -\,0.1742 \\ \hline 0.8858 \end{array}$$ In this case, it is necessary to use zeros as place holders. This gives us the ability to borrow and solve the subtraction.

Problems   Line up vertically according to place values, then subtract.

1) $48.6 - 21.831 =$ _____

3) $8.999 - 7.983 =$ _____

2) $6 - 3.91 =$ _____

4) $5.52 - 4.116 =$ _____

5) $8963.8 - 2.0004 =$ _____

8) $4.44 - 0.06 =$ _____

6) $23.4 - 23.04 =$ _____

9) $1753.8 - 9.9 =$ _____

7) $100 - 99.99 =$ _____

10) $652.9 - 651.667 =$ _____

11) Steve Morrison made a deposit of $1640.00 into a savings account. On March 6, he withdrew $110.87. What is his new balance?   _____

12) RaeAnn set up an IV with 2.5 l of IV solution for a patient. When she returned, the bag contained only 1.755 l. How much had been absorbed?   _____

13) Larry Compton is an anesthesiologist. He administers Vidalar to patients undergoing oral surgery. The following chart is used:

Patient's weight (lb)	Mg given
50–75	28.5
75–100	30.3
100–150	38.2
150–200	45.6
200+	49.0

a. How much greater is the amount of Vidalar given to a 110-pound patient than to an 85-pound patient?   _____

b. Larry administers Vidalar gradually through IV flow. If a 140-pound patient has received 32.068 mg, how much more does he need?   _____

c. If a patient is right on the border of a weight category, Larry finds the difference between the two categories' dosages. What is the difference in dosages for the two categories that a 200-pound man fits in?   _____

14) Find the difference in the following two list prices.

file folders (Aplar brand) $4.528/doz
file folders (Jaguar brand)   4.67/doz   _____

15) Jeremy Bartmond works at Walten County Home as a dietician. He has the recipes he uses on computer. All fractions are entered as decimals (rounded to 5 places). Find the difference in salt used in the following two recipes.   _____

Stroganoff
25 lb chopped meat
1 gal mushroom soup
18 lb noodles
1.625 tsp salt
3 gal sour cream

Stroganoff 2
20 lb chopped meat
1 gal sour cream
20 lb noodles
4.5 c flour
1 gal beef stock
3.33333 tsp salt

Multiplication

To multiply decimal numbers, it is not necessary to line up the place values. Instead, write the numbers so that they align on the right.

Example 1

19.6×2.01

$$\begin{array}{r} 19.6 \\ \times 2.01 \\ \hline \end{array}$$

The process of multiplication is the same as in whole number multiplication except that when you have completed the multiplication, it is necessary to place the decimal in your answer.

$$\begin{array}{r} 19.6 \\ \times 2.01 \\ \hline 196 \\ 000 \\ 392 \\ \hline 39396 \end{array}$$
$$\begin{array}{r} 19.\mathbf{6} \\ \times 2.\mathbf{01} \\ \hline 196 \\ 000 \\ 392 \\ \hline 39.\mathbf{396} \end{array}$$

A total of 3 numbers are to the right of the decimal—

and there should be the same number in the answer.

Problems   Line up vertically, then multiply and place the decimal in the answer.

1) $6.007 \times .001 =$ _____

2) $9.72 \times 14.6 =$ _____

3) $48.64 \times 5.3 =$ _____

4) $9.61 \times 8.14 =$ _____

5) $8.081 \times .005 =$ _____

6) $9.9 \times 8.76 =$ _____

7) $1463 \times 5.7 =$ _____

8) $59.232 \times 14.8 =$ _____

9) $11.652 \times 7.7 =$ _____

10) $436.49 \times 2.117 =$ _____

11) Shane Fayler earns $33.75 per visit as an elderly assistance counselor. He made 55 visits during August. What is his total pay? _____

12) To find the area of a room, it is necessary to multiply length by width (see figure 3–1). Jack Hillburn needs to find the area of his office to order floor covering. What is the area of his office (with the following dimensions: $18.5' \times 12.625'$? _____

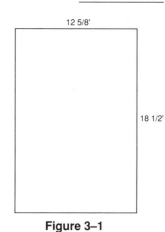

12 5/8'

18 1/2'

Figure 3–1

13) Kayla Hardesty sells home health supplies. Complete the following invoice.

Quantity	Item	Unit Cost	Extension
2	179–2 mattress cover	$ 17.489	_____
1	9#3 daybed	896.51	_____
3	473–6 gloves	7.80	_____

14) Angela Gordensen has an IV with a flow rate of 1.75 ml per hour. If her IV is in for 6.5 hours, what is the total intake (in ml)? _____

15) Johnna Keelor supervises 3–F Nursing Station. At the end of each week, she determines the total hours worked for each person and reports them to Personnel.

Employee #	Days	Length of Shift (hr)	Total
17	5	8.5	_____
11	5	6.75	_____
13	4	10	_____
14	6	5.25	_____

Division

To divide decimal numbers, it is necessary to convert the divisor (number being divided by) to a whole number. To do so, you can move the decimal to the right until it is at the right of the number.

Example 1

$1954.66 \div 5.67 =$

1. First, set up the problem under a division sign.

$$5.67\overline{)1954.66}$$

2. Then, move the decimal in the divisor (5.67) to the right.

$$5.67\overline{)1954.66} \qquad \text{becomes} \qquad 567\overline{)1954.66}$$

Changing 5.67 to 567 is accomplished by moving the decimal, but it is actually the process of multiplying 5.67 by 100 ($5.67 \times 100 = 567$). This same multiplication must be done on the dividend (number being divided) for the result to be correct.

3. Next, change the dividend by multiplying (by 100 in this case).

$$1954.66 \times 100 = 195466$$

This is the same as moving the decimal the same number of places in the dividend as you did in the divisor.

$$567\overline{)1954.66} \qquad \text{becomes} \qquad 567\overline{)195466}$$

4. Once the decimal has been moved out of the divisor and moved by similar multiplication in the dividend, division is completed as in whole number division.

$$
\begin{array}{r}
344.737 \\
567.\overline{)195466.} \\
\end{array}
$$

$$
\begin{array}{r}
-1701 \\
\hline
2536 \\
-2268 \\
\hline
2686 \\
-2268 \\
\hline
4180 \\
-3969 \\
\hline
2110 \\
-1701 \\
\hline
4090 \\
-3969 \\
\hline
121 \\
\end{array}
$$

Division is completed to the thousandths place so that rounding can be done to the hundredths place.

Zero place holder.

Zero place holder.

Zero place holder.

Answer: 344.74 after rounding

Instead of having a remainder if the division does not come out evenly, it is common to continue division with the use of zero place holders. If it does not come out evenly then, it will be necessary to round the answer to some designated place value. If a place value is specified in the problem, follow that guideline. If not, it is acceptable to round the answer to the hundredths place.

Problems   Remove the decimal from the divisor by moving it to the right. Then, move the decimal in the dividend in the same manner (by moving the same number of decimal places as in divisor). Then, divide as in whole number division and round to the hundredths place.

1) $36.22\overline{)56.374}$ _____

2) $264.1\overline{)784.4}$ _____

3) $5.1\overline{)1201.05}$ _____

4) $4.6\overline{)741.4}$ _____

5) $223.4\overline{)1797}$ _____

6) $4.2\overline{)2893.8}$ _____

7) $80.5\overline{)336.1}$ _____

8) $78.4\overline{)163.2}$ _____

9) $3.6\overline{)82.4}$ _____

10) $486.325\overline{)3755.8}$ _____

11) Greg Cassidy works at Reese Pharmacy as a pharmacist. He filled an order for 26.5 mg of Conbetrex. If the stock bottle contains 100 mg, how many equal 26.5 mg prescriptions could he fill from that bottle?   _____

12) Helen Jackson has 45.8 pounds to lose, according to her physician. She is supposed to lose it gradually over a ten-month period. How many pounds should she try to lose each month?   _____

13) Jennifer works at Clinic Dieticians as a nutrition consultant to the elderly. Recently, she advised Mr. Thomas to control his sodium intake. One food that he asked about eating was CrackleWheat Crackers. Each cracker contains 27.8 mg of sodium. If he should have only 300 mg on that particular snack, how many crackers can he have? _____

14) Delores Miller needs to find the average of her blood sugar counts over the last 16 days. The total of the blood sugar readings is 3433.6. What is her average blood sugar count? _____

15) Rosa Nuñez went shopping for her invalid mother. She compared two brands of peanut butter. Brand A costs $5.67 for 25 ounces and Brand B costs $6.14 for 28 ounces. To find the better buy, she first must find the unit cost of each item. To do this, divide the cost by the number of ounces.

   a. What is the unit cost of Brand A per ounce? _____

   b. What is the unit cost of Brand B per ounce? _____

   c. Which is cheaper? _____

Often, the unit cost of an item will be displayed on the grocer's shelf to aid in comparison. The unit pricing label for an item is shown (see figure 3–2).

Figure 3–2   Label indicates price per ounce—unit price.
(Courtesy of Clyde Evans Supermarket)

Mixed Operations

Problems   Complete the following decimal operations problems by first determining what operations must be done and then solving the problem. More than one step may be required.

1) Shannon Brown was asked to find the average weight of the supply shipments being sent from her department at Weger Hospital Supply Company. Find the average of the following.

576.33	678.33	35.2
26.76	15	32.87
1.7	1,467.87	26.8
347	582.49	485.05

              _____

2) Tim Normandy bought a new uniform from Halsey Uniforms for $26.79 plus tax. If sales tax is found by multiplying the cost of the items by .065, what is the sales tax? _____

What is the total cost of the uniform, including tax? _____

3) Brian Hudson heard that the nursing staff at Golliver Hospital earns $13.57 per hour. He is currently earning $12.66 per hour. How much more could he earn at Golliver than at his current job? _____

4) Natalie Greenwood drives 5.7 miles to work each day (one-way trip). Gasoline costs $1.09 per gallon, and she can drive 34.2 miles on a gallon of gas. How much does one day's drive to and from work cost her for gasoline? _____

5) Raphael Bonnare orders supplies for his home-based x-ray service. He compares the cost of x-ray film from two different companies. Gladwell, Inc., sells x-ray film for $77.39 for 25 exposures. Frampton X-ray sells the film for $100.99 for 32 exposures.

   a. What is the unit cost of Gladwell's film per exposure? _____

   b. What is the unit cost of Frampton's film per exposure? _____

   c. Which one is less expensive and by how much? _____

6) Phil bought four textbooks at University Bookstore. His total bill was $57.92. What was the average cost of the books? _____

7) Joanne drives 17.8 miles to work, 5.2 miles from there to her evening college class, and 16.2 miles from there back home. What is her total driving distance each day? _____

8) Cherie Preston used a 28.25-ounce can of kidney beans in a recipe. If each portion gets an equal amount of beans and the recipe serves three people, how many ounces of beans will each serving contain? _____

9) Daniel Douglas found a sale on the stethoscope that he needed to buy. The original cost was $69.75, and the sale cost was $57.98. How much could he save by buying the stethoscope on sale? _____

10) Sarah weighed 7.95 kilograms at her six-month checkup. If each kilogram is equal to 2.2 pounds, how many pounds did she weigh? _____

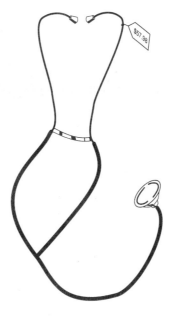

Figure 3–3

Chapter 4

Fraction Operations

Fractions are used in a wide variety of ways in the health field. Medications may be prescribed as a fraction. Diet and nutrition information can contain fractions. Simple office procedures such as payroll, ordering, or bill collection often contain fractions. The more familiar you are with fraction operations, the better equipped you will be for any job in the health field.

Defining a Fraction

What is a fraction? A fraction is a symbolic way of writing a **part** of something. Just as the decimal $.65 shows **part** of a dollar, so $\frac{7}{8}$ pound shows **part** of a pound (see figure 4–1).

A fraction can be written in two different ways: $\frac{7}{8}$ or 7/8.

A fraction is made up of two parts, the numerator and the denominator. The numerator is the top number (or first number), and it signifies how many parts of the whole are being represented by the fraction. The denominator is the bottom number (or second number), and it signifies how many parts the whole has been divided or sectioned into.

For example, in $\frac{7}{8}$ the number 7 is the numerator and 8 is the denominator. The fraction $\frac{7}{8}$ means that 7 parts out of 8 possible are being represented. If you have an order containing 8 cases of medical supplies and 7 have arrived, you have 7 cases out of the 8 that could possibly be sent. You have $\frac{7}{8}$ of the order.

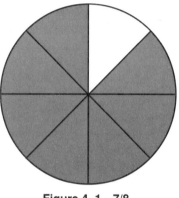

Figure 4–1   7/8

Another example is in money. If you have only 1 quarter, you have 1 quarter out of the 4 that you would need to equal a dollar. The fraction to represent this is 1/4 or $\frac{1}{4}$. You have one-fourth ($\frac{1}{4}$) a dollar.

Or suppose that you were supposed to administer 20 mg of Solucene to a patient, and each Solucene tablet contained 40 mg. You need 20 mg out of 40 mg available, or $\frac{20}{40}$. A fraction can be reduced if the numerator and denominator have a number in common that will equally divide into each—in this case, 20 will divide into 20 and also into 40.

$$\frac{20}{40} \div \frac{20}{20} = \frac{1}{2}$$

You would administer $\frac{1}{2}$ a tablet to the patient (see figure 4–2).

Figure 4–2   Most pills can be split to administer half a pill.

Least Common Denominators

To compare two fractions, it is necessary that both fractions have the same denominator. Remember that the denominator shows how many parts the whole has been divided or sectioned into. Once the two fractions have the same denominator, it is possible to look at the numerators to decide which fraction is larger. For example, $\frac{5}{8}$ and $\frac{3}{8}$ have the same denominator. Let these fractions represent a nurse's work shift. The first fraction indicates that the nurse has worked 5 hours out of 8 possible. The second fraction indicates that the nurse has worked 3 hours out of 8 possible. Which time period is longer, 5 hours or 3 hours? It is easy to see that 5 hours is the larger time period. But, if the shifts had not both been equally divided into eight one-hour periods, comparison would have been more difficult.

It is possible to create like denominators for fractions. The process of doing so is called finding the **least common denominator**. A least common denominator between two or more fractions is the smallest number possible that all of the available denominators can divide into. For example, in the fractions $\frac{1}{2}$ and $\frac{3}{5}$, 2 and 5 are the denominators. Since they are not like denominators (the same number), we must find the least common denominator. The question we must ask is: what is the smallest number that 2 and 5 can both divide into?

Multiples of 2: 2   4   6   8   **10**

Multiples of 5: 5   **10**

Ten is the smallest number that 2 and 5 will both divide into. Therefore, 10 is the least common denominator for $\frac{1}{2}$ and $\frac{3}{5}$.

Now, to use the least common denominator (LCD) to help us compare the fractions $\frac{1}{2}$ and $\frac{3}{5}$, we must make 10 the new denominator in both fractions.

$\frac{1}{2} = \frac{?}{10}$ We cannot just call this $\frac{1}{10}$ now. That is not equal to $\frac{1}{2}$. We need to find an equal fraction. **Equal fractions** are fractions where the numerator and denominator of one fraction can be multiplied by a number to result in the second fraction.

$\frac{1}{2} = \frac{5}{10}$ 2 times another number = 10. Whatever number you must multiply by to result in 10, multiply the 1 times that same number to find the new numerator.

2 times 5 = 10. So, multiply $1 \times 5 = 5$. These are equal fractions.

$\frac{3}{5} = \frac{?}{10}$ We cannot just call this $\frac{3}{10}$ now. That is not equal to $\frac{3}{5}$.

5 times another number = 10. Whatever number you must multiply by to result in 10, multiply the 3 times that same number to find the new numerator.

$\frac{3}{5} = \frac{6}{10}$ 5 times 2 = 10. So, multiply $3 \times 2 = 6$. These are equal fractions.

Now you can compare $\frac{1}{2}$ and $\frac{3}{5}$. Look at their equal fractions ($\frac{1}{2} = \frac{5}{10}$ and $\frac{3}{5} = \frac{6}{10}$). $\frac{5}{10}$ and $\frac{6}{10}$ can be compared easily. Since 6 parts is more than 5 parts, $\frac{6}{10}$ is larger. Therefore, $\frac{3}{5}$ is larger than $\frac{1}{2}$.

Finding the least common denominator of two or more fractions is the key to being able to compare. Having the LCD is also necessary in addition or subtraction of fractions. Let's look more closely at how to find the LCD.

Example 1

Find the LCD of $\frac{4}{6}$ and $\frac{7}{9}$.

1. You are trying to find the smallest number that the denominators, 6 and 9, will both divide evenly into.

2. Look at the multiples of 6 and of 9 until you find a number in common. (Finding the multiples of a number is the same as counting by that number.)

    Multiples of 6:   6   12   **18**
    Multiples of 9:   9   **18**

3. 18 is the least common denominator.

Example 2

Find the LCD of $\frac{3}{7}$ and $\frac{5}{8}$.

    Multiples of 7:   7   14   21   28   35   42   49   **56**
    Multiples of 8:   8   16   24   32   40   48   **56**

The least common denominator of $\frac{3}{7}$ and $\frac{5}{8}$ is 56 .

Example 3

Find the LCD of $\frac{1}{12}$ and $\frac{5}{6}$.

    Multiples of 12:   **12**
    Multiples of 6:   6   **12**

The least common denominator of $\frac{1}{12}$ and $\frac{5}{6}$ is 12.

Example 4

Find the LCD of $\frac{1}{6}$ and $\frac{3}{8}$, then convert each fraction to an equal fraction using the LCD.

1. Find the LCD.

    Multiples of 6:   6   12   18   **24**
    Multiples of 8:   8   16   **24**

2. Make the LCD the new denominator in each fraction and determine the new numerator for each by multiplication.

$$\frac{1}{6} \rightarrow \frac{}{24} \qquad \begin{array}{l} 6 \times \,? = 24 \\ 6 \times 4 = 24 \end{array}$$

So, multiply 1×4 to get 4 as the new numerator.

$$\frac{1}{6} \rightarrow \frac{4}{24}$$

$$\frac{3}{8} \rightarrow \frac{}{24} \qquad \begin{array}{l} 8 \times \,? = 24 \\ 8 \times 3 = 24 \end{array}$$

So, multiply 3×3 to get 9 as the new numerator.

$$\frac{3}{8} \rightarrow \frac{9}{24}$$

The LCD is 24, and the new fractions are $\frac{1}{6} = \frac{4}{24}$ and $\frac{3}{8} = \frac{9}{24}$.

Example 5

Find the LCD for $\frac{4}{5}$, $\frac{3}{9}$, and $\frac{5}{6}$.

Multiples of 5: 5   10   15   20   25   30   35   40   45   50
Multiples of 9: 9   18   27   36   45   54   63   72   81   90
Multiples of 6: 6   12   18   24   30   36   42   48   54   60

After a considerable list, an LCD still is not found. Be sure you haven't made a mistake and keep trying.

Multiples of 5: 5   10   15   20   25   30   35   40   45   50
                55   60   65   70   75   80   85   **90**   95
Multiples of 9: 9   18   27   36   45   54   63   72   81   **90**
Multiples of 6: 6   12   18   24   30   36   42   48   54   60
                66   72   78   84   **90**   96

The LCD is 90.

Problems   Find the LCD for the following pairs or groups of fractions.

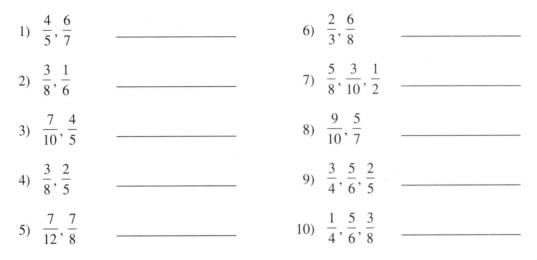

1) $\frac{4}{5}, \frac{6}{7}$ _____

6) $\frac{2}{3}, \frac{6}{8}$ _____

2) $\frac{3}{8}, \frac{1}{6}$ _____

7) $\frac{5}{8}, \frac{3}{10}, \frac{1}{2}$ _____

3) $\frac{7}{10}, \frac{4}{5}$ _____

8) $\frac{9}{10}, \frac{5}{7}$ _____

4) $\frac{3}{8}, \frac{2}{5}$ _____

9) $\frac{3}{4}, \frac{5}{6}, \frac{2}{5}$ _____

5) $\frac{7}{12}, \frac{7}{8}$ _____

10) $\frac{1}{4}, \frac{5}{6}, \frac{3}{8}$ _____

In problems 11–15, find the LCD, then convert each fraction to an equal fraction using the LCD.

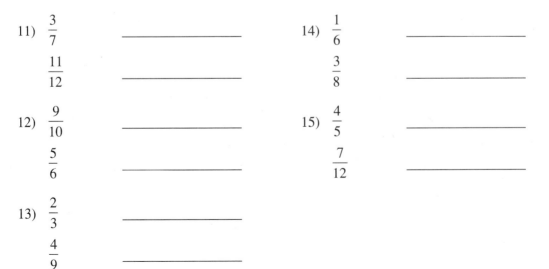

11) $\frac{3}{7}$ _____

14) $\frac{1}{6}$ _____

$\frac{11}{12}$ _____

$\frac{3}{8}$ _____

12) $\frac{9}{10}$ _____

15) $\frac{4}{5}$ _____

$\frac{5}{6}$ _____

$\frac{7}{12}$ _____

13) $\frac{2}{3}$ _____

$\frac{4}{9}$ _____

In problems 16–20, compare the fractions and determine which is largest.

16) $\dfrac{5}{7}, \dfrac{3}{4}$ _____

18) $\dfrac{12}{15}, \dfrac{9}{10}$ _____

20) $\dfrac{3}{5}, \dfrac{4}{6}$ _____

17) $\dfrac{6}{9}, \dfrac{5}{6}$ _____

19) $\dfrac{5}{6}, \dfrac{4}{7}$ _____

Reducing Fractions

Often in working with fractions, it is necessary to reduce fractions. Reducing a fraction involves dividing the numerator and denominator both by the largest number that will divide evenly into both (**greatest common factor: GCF**). For example, in $\frac{10}{40}$ the factors of 10 are 1, 2, 5, and 10, and the factors of 40 are 1, 2, 4, 5, 8, and 10. The factors that 10 and 40 have in common are: 2, 5, and 10. Which of these is the greatest? Ten is the greatest common factor. Thus, this fraction can be reduced by dividing the numerator and denominator both by 10 (see figure 4–3).

$$\frac{10}{40} \rightarrow \frac{10}{10} = \frac{1}{4}$$

Another method of reducing numbers does not require finding the greatest common factor. To reduce a fraction without finding the GCF, simply divide the numerator and denominator by any number that will divide evenly into both. Continue to do this until there are no other numbers that will divide evenly into both. For example, in the $\frac{10}{40}$ problem, you could:

$$\frac{10}{40} \rightarrow \frac{2}{2} = \frac{5}{20} \rightarrow \frac{5}{5} = \frac{1}{4}$$

There are more steps, but this method eliminates the need to find the GCF.

Problems Reduce the following fractions to lowest terms (meaning that there are no other numbers that will divide evenly into the numerator and denominator when you are done).

1) $\dfrac{6}{48}$ _____

3) $\dfrac{5}{20}$ _____

5) $\dfrac{9}{54}$ _____

2) $\dfrac{3}{9}$ _____

4) $\dfrac{8}{12}$ _____

6) $\dfrac{14}{21}$ _____

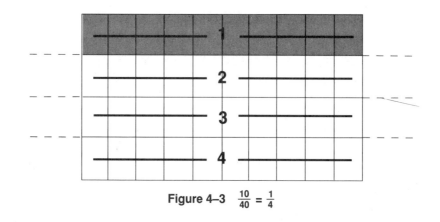

Figure 4–3 $\frac{10}{40} = \frac{1}{4}$

7) $\frac{6}{21}$ _____ 12) $\frac{9}{24}$ _____ 17) $\frac{4}{18}$ _____

8) $\frac{9}{15}$ _____ 13) $\frac{10}{34}$ _____ 18) $\frac{3}{42}$ _____

9) $\frac{12}{18}$ _____ 14) $\frac{15}{33}$ _____ 19) $\frac{2}{16}$ _____

10) $\frac{8}{18}$ _____ 15) $\frac{9}{12}$ _____ 20) $\frac{7}{28}$ _____

11) $\frac{8}{20}$ _____ 16) $\frac{8}{14}$ _____

Mixed Numbers and Improper Fractions

A **mixed number** is a combination of a whole number and a fraction written together. For example, Joyce got $3\frac{1}{2}$ books read from the assignment. Or, Steve recorded $4\frac{1}{5}$ patient records before his break. In each of these examples, the whole number indicates the number of complete (whole) activities accomplished and the fraction indicates that part, but not a whole, of another activity was accomplished.

An **improper fraction** is a fraction where the numerator is larger than the denominator. For example, the number $\frac{7}{2}$ is an improper fraction, as is $\frac{21}{5}$.

When writing numbers in lowest terms, it is necessary to change any improper fractions to a mixed number. This is a process of dividing the denominator into the numerator to find out how many **whole** groups there are—this would be the whole number. Then, the remainder that is left after division, if any, is the number of **parts** left, and this becomes the numerator of the fraction. For example, in $\frac{7}{2}$ we will first divide 7 by 2.

$$\begin{array}{r} 3 \\ 2\overline{)7} \\ -6 \\ \hline 1 \end{array}$$

The result is the whole number 3. There are three **whole** groups of 2 in the number 7. The remainder is 1. The **part** that is left is 1 out of 2 possible, or $\frac{1}{2}$. The improper fraction $\frac{7}{2}$ becomes $3\frac{1}{2}$. There are 3 groups of 2 in 7 and there is 1 left over—1 part of a group of 2. Thus, $3\frac{1}{2}$ (see figure 4–4).

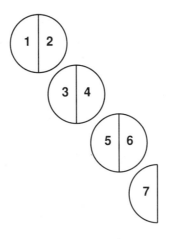

Figure 4–4 7 halves equals 3 wholes and 1 half. $\frac{7}{2} = 3\frac{1}{2}$.

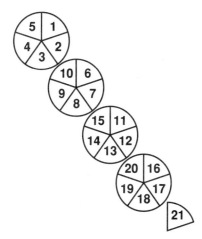

Figure 4–5 4 wholes and 1 fifth equals 21 fifths. $4\frac{1}{5} = \frac{21}{5}$

Sometimes it is necessary to change a mixed number back to an improper fraction. This is necessary when working with some fraction operations, such as multiplication and division. The procedure for changing a mixed number to an improper fraction is the opposite of what you do in division above. For division, you divide to see how many times 2 will go into 7 equally, then you subtract to find the remainder. To "undo" this, first you will multiply the denominator (2) by the whole number (3) to get 6 total parts (3 groups of 2). Then, you will add the extra part (the remainder 1) on to the 6 (6 + 1 = 7). You have 7 parts that should be in groups of 2. Thus, the improper fraction $\frac{7}{2}$.

For another example of this, let's look at $4\frac{1}{5}$. You have 4 whole groups of 5 (4 × 5 = 20). Add the extra part (the numerator 1) on to the 20 (20 + 1 = 21). You have 21 parts that should be in groups of 5. Thus, the improper fraction $\frac{21}{5}$ (see figure 4–5).

Problems Change the following mixed numbers to improper fractions.

1) $4\frac{2}{7}$ _____   6) $3\frac{1}{6}$ _____   11) $4\frac{1}{3}$ _____   16) $7\frac{2}{5}$ _____

2) $1\frac{1}{2}$ _____   7) $6\frac{2}{3}$ _____   12) $6\frac{1}{4}$ _____   17) $4\frac{3}{4}$ _____

3) $5\frac{2}{6}$ _____   8) $4\frac{1}{7}$ _____   13) $3\frac{5}{6}$ _____   18) $1\frac{1}{6}$ _____

4) $6\frac{2}{3}$ _____   9) $10\frac{3}{8}$ _____   14) $5\frac{1}{5}$ _____   19) $4\frac{3}{7}$ _____

5) $3\frac{1}{4}$ _____   10) $3\frac{7}{10}$ _____   15) $2\frac{7}{9}$ _____   20) $3\frac{1}{3}$ _____

In problems 21–40, change the following improper fractions to mixed numbers in lowest terms.

21) $\frac{15}{4}$ _____   26) $\frac{19}{3}$ _____   31) $\frac{16}{3}$ _____   36) $\frac{19}{2}$ _____

22) $\frac{18}{14}$ _____   27) $\frac{16}{7}$ _____   32) $\frac{25}{4}$ _____   37) $\frac{18}{5}$ _____

23) $\frac{22}{5}$ _____   28) $\frac{31}{6}$ _____   33) $\frac{40}{3}$ _____   38) $\frac{38}{4}$ _____

24) $\frac{5}{2}$ _____   29) $\frac{14}{7}$ _____   34) $\frac{53}{20}$ _____   39) $\frac{45}{2}$ _____

25) $\frac{42}{5}$ _____   30) $\frac{6}{4}$ _____   35) $\frac{55}{6}$ _____   40) $\frac{23}{7}$ _____

Addition

To add fractions, it is first necessary to have like denominators (as we used in comparing fractions). Then, once the denominators are the same, you can add the numerators together to find the total number of parts in the whole. Occasionally, the total in the numerator will exceed the number in the denominator. When this occurs, you have an improper fraction. This must then be written in lowest terms as a mixed number.

Example 1

Add $\frac{5}{6} + \frac{2}{15}$

$$\frac{5}{6} \quad \frac{25}{30}$$ First, change the denominators to the LCD.

$$+\frac{2}{15} \quad \frac{4}{30}$$ Then, multiply to find the new numerators.

$$\frac{29}{30}$$ Next, add the numerators together and leave the denominator alone. Reduce if possible.

Example 2

Add $\frac{4}{5} + \frac{7}{8}$

$$\frac{4}{5} \quad \frac{32}{40}$$ First, change the denominators to the LCD.

$$+\frac{7}{8} \quad \frac{35}{40}$$ Then, multiply to find the new numerators.

$$\frac{67}{40}$$ Next, add the numerators together and leave the denominator alone. Reduce this fraction because it is an improper fraction ($67 \div 40 = 1$ remainder 27).

$$1\frac{27}{40}$$

Sometimes adding fractions includes adding mixed numbers as well, either with another mixed number or with a proper fraction. The addition process for the fractions is the same. If the answer is an improper fraction, it still needs to be converted to a mixed number first. Then the whole number which that creates is carried over and added with the other whole numbers.

Example 3

Add $4\frac{7}{8} + 2\frac{3}{4}$

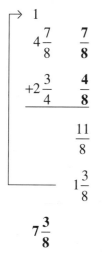

$$4\frac{7}{8} \quad \frac{7}{8}$$

First, change the denominators to the LCD.

$$+2\frac{3}{4} \quad \frac{4}{8}$$

Then, multiply to find the new numerators.

$$\frac{11}{8}$$

Next, add the numerators together and leave the denominator alone. Reduce this fraction because it is an improper fraction. ($11 \div 8 = 1$ remainder 3)

$$1\frac{3}{8}$$

$$7\frac{3}{8}$$

Carry the whole number 1 over to the whole numbers and add all whole numbers together. $7\frac{3}{8}$ is the final answer. The arrow indicates carrying.

Problems Add the following pairs or groups of fractions, and write the final answer in lowest terms.

1) $\quad 3\frac{6}{10}$
   $\quad + \ 5\frac{1}{3}$

2) $\quad \frac{14}{15}$
   $\quad + \ \frac{4}{6}$

3) $\quad \frac{12}{14}$
   $\quad + \ 7\frac{7}{10}$

4) $\quad \frac{9}{14}$
   $\quad + \ 14\frac{13}{21}$

5) $\quad \frac{8}{9}$
   $\quad + \ 2\frac{7}{12}$

6) $\quad \frac{11}{12}$
   $\quad + \ \frac{8}{15}$

7) $\quad \frac{3}{10}$
   $\quad + \ \frac{3}{4}$

8) $\quad \frac{10}{12}$
   $\quad + \ \frac{5}{21}$

9) $\quad 7\frac{4}{5}$
   $\quad + \ \frac{5}{8}$

10) $\quad \frac{13}{14}$
    $\quad + \ 8\frac{4}{7}$

Subtraction

To subtract fractions, it is first necessary to have like denominators (as we used in comparing fractions and in addition of fractions). Then, once the denominators are the same, you can subtract the numerators to find the difference. Occasionally, the two numerators cannot be subtracted because the bottom one is larger than the top one from which it is being subtracted. When this occurs, you must borrow from the whole number on top.

Borrowing in fractions is quite a bit different from borrowing in whole numbers. In regular whole number subtraction, you borrow from the tens column to increase the value of the ones column. For example:

$$\begin{array}{r} 1\ 15 \\ \cancel{2}5 \\ -18 \\ \hline \end{array}$$

Borrowing 1 from the 2 in the tens column allowed you to increase the 5 in the ones column to 15 (which is 5 plus the 10 borrowed).

In fraction borrowing, however, you borrow from a whole number to increase the numerator of the top fraction. When you borrow 1 from the whole number, you are really borrowing all of the parts of 1 whole unit. For example, you might be borrowing 4 parts of a unit that is sectioned into fourths or 7 parts of a unit that is sectioned into sevenths. You are borrowing a whole unit regardless of how it is sectioned or divided up.

Example 1

$$\cancel{5}\frac{\cancel{1}}{6} \qquad 4\frac{7}{6}$$
$$-2\frac{5}{6}$$
$$\overline{\quad 2\frac{2}{6}\quad}$$

First, recognize that $\frac{1}{6} - \frac{5}{6}$ cannot be subtracted.

Next, borrow 1 from the 5 and change that one that is borrowed to 6 parts of a unit that is sectioned into sixths. (Why sixths? Because that is the common denominator of the problem.)

Then add the $\frac{6}{6}$ (six sixths borrowed) to the $\frac{1}{6}$ already available ($\frac{6}{6} + \frac{1}{6} = \frac{7}{6}$).

$\frac{7}{6}$ is large enough to complete the subtraction problem ($\frac{7}{6} - \frac{5}{6} = \frac{2}{6}$).

Reduce the fraction if necessary.

Final answer: $2\frac{1}{3}$.

Be sure when subtracting fractions that you put the problem in like denominators first before deciding whether or not to borrow.

Example 2

$$3\frac{1}{9}$$
$$-1\frac{3}{4}$$
$$\overline{\qquad}$$
$$\cancel{3}\frac{\cancel{4}}{36} \qquad 2\frac{40}{36}$$
$$-1\frac{27}{36}$$
$$\overline{\quad 1\frac{13}{16}\quad}$$

First, rewrite the problem with like denominators (LCD for 9 and 4 is 36).

$\frac{1}{9}$ becomes $\frac{4}{36}$.

$\frac{3}{4}$ becomes $\frac{27}{36}$.

Now, realizing that you cannot subtract $\frac{4}{36}$ from $\frac{27}{36}$, you must borrow.

Borrow 1 from the 3 and consider it $\frac{36}{36}$.

Add the $\frac{36}{36}$ that was borrowed to the $\frac{4}{36}$ already available ($\frac{36}{36} + \frac{4}{36} = \frac{40}{36}$).

Now you are able to subtract. ($\frac{40}{36} - \frac{27}{36} = \frac{13}{36}$ and $2 - 1 = 1$ in the whole numbers.)

Reduce if necessary.

Final answer: $1\frac{13}{36}$.

Problems   Subtract the following pairs of fractions and write the answer in lowest terms.

1)	$\frac{5}{8}$ $-\frac{4}{8}$	6)	$\frac{8}{10}$ $-\frac{3}{4}$

1) $\frac{5}{8}$ 6) $\frac{8}{10}$ 11) $7\frac{3}{10}$
         $-\frac{4}{8}$ $-\frac{3}{4}$ $-4\frac{4}{5}$

2) $\frac{14}{15}$ 7) $\frac{3}{10}$ 12) $2\frac{8}{9}$
         $-\frac{4}{6}$ $-\frac{1}{10}$ $-2\frac{7}{12}$

3) $\frac{12}{14}$ 8) $\frac{13}{14}$ 13) $4\frac{7}{9}$
         $-\frac{7}{10}$ $-\frac{4}{7}$ $-1\frac{1}{3}$

4) $\frac{11}{12}$ 9) $\frac{7}{12}$ 14) $7\frac{4}{5}$
         $-\frac{5}{12}$ $-\frac{3}{8}$ $-\frac{5}{8}$

5) $\frac{11}{12}$ 10) $7\frac{6}{10}$ 15) $16\frac{2}{5}$
         $-\frac{8}{15}$ $-5\frac{1}{3}$ $-13\frac{7}{10}$

Multiplication

Multiplying fractions is the same as trying to find a certain part of a part. For example, what is $\frac{1}{2}$ of $\frac{3}{4}$? This is a multiplication problem. To understand what $\frac{1}{2}$ of $\frac{3}{4}$ really is, let's look at the following diagram (figure 4–6):

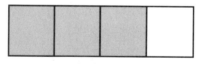

Figure 4–6

At first, $\frac{3}{4}$ of the diagram was filled in (indicated by the first three boxes). Then, in finding $\frac{1}{2}$ of that, half of each box was shaded in. Half of $\frac{3}{4}$ is represented by the shaded part. This is really equivalent to 3 parts out of 8 possible (or $\frac{3}{8}$), as shown by the next diagram (figure 4–7):

Figure 4–7   **Three shaded areas out of eight possible, or $\frac{3}{8}$.**

When you multiply fractions, it is not necessary to have like denominators. It is important, however, that all mixed numbers are changed to improper fractions first before multiplying.

Then, you multiply the numerators together and write the result on top, and multiply the denominators together and write the result on the bottom. Then, reduce to the lowest terms if necessary.

$$\frac{1}{2}\times\frac{3}{4}=\frac{3}{8} \qquad \begin{array}{l} -1\times3=3 \\ -2\times4=8 \end{array}$$

Example 1

Multiply $\frac{4}{5} \times \frac{7}{8}$

$$\frac{4}{5} \times \frac{7}{8} = \frac{28}{40} \qquad \begin{array}{l} -\ 4 \times 7 = 28 \\ -\ 5 \times 8 = 40 \end{array}$$

This can be reduced, since 4 divides into 28 and 40 evenly.

$$\frac{28}{40} \div \frac{4}{4} = \frac{7}{10} \qquad \text{The final answer is } \tfrac{7}{10}.$$

It is also possible to reduce before you multiply. If either numerator and either denominator can be divided evenly by some number, that can be done first before multiplication. In example 1, the 4 in $\frac{4}{5}$ (a numerator) and the 8 in $\frac{7}{8}$ (a denominator) both can be divided evenly by 4. Do this first and see what the result will be.

$$\frac{\overset{1}{4}}{5} \times \frac{7}{\underset{2}{8}} = \frac{7}{10} \qquad \text{You still get } \tfrac{7}{10} \text{ as your final answer.}$$

Example 2

Multiply $1\frac{3}{4} \times 2\frac{5}{6}$

$$1\frac{3}{4} \times 2\frac{5}{6} =$$

You must change all mixed numbers to improper fractions first.

$$\frac{7}{4} \times \frac{17}{6} =$$

Now you can multiply the numerators and the denominators or reduce first, if possible.

$$\frac{7}{4} \times \frac{17}{6} = \frac{119}{24}$$

This must then be reduced to lowest terms (that is, back to a mixed number in lowest terms).

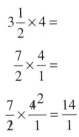

The final answer is $4\frac{23}{24}$.

Example 3

Multiply $3\frac{1}{2} \times 4$

$$3\frac{1}{2} \times 4 =$$

You must change all mixed numbers to improper fractions first (the number 4 becomes 4 over 1).

$$\frac{7}{2} \times \frac{4}{1} =$$

Now you can multiply the numerators and the denominators or reduce first, if possible.

$$\frac{7}{2} \times \frac{\overset{2}{4}}{1} = \frac{14}{1}$$

This fraction equals 14 whole units, or 14.

Problems   Multiply the following pairs of fractions or mixed numbers and reduce to lowest terms.

1) $1\frac{2}{3} \times \frac{5}{6}$ _____

2) $\frac{2}{5} \times \frac{7}{8}$ _____

3) $\frac{7}{10} \times \frac{3}{4}$ _____

4) $\frac{1}{2} \times \frac{4}{5}$ _____

5) $1\frac{4}{9} \times \frac{3}{4}$ _____

6) $3\frac{1}{4} \times 2\frac{8}{9}$ _____

7) $4\frac{1}{6} \times \frac{3}{8}$ _____

8) $3\frac{3}{4} \times \frac{4}{5}$ _____

9) $1\frac{4}{7} \times \frac{7}{8}$ _____

10) $4\frac{1}{5} \times 6\frac{5}{8}$ _____

11) $5\frac{2}{5} \times \frac{7}{10}$ _____

12) $2\frac{1}{4} \times 4\frac{5}{6}$ _____

13) $7\frac{2}{3} \times 3\frac{3}{7}$ _____

14) $3\frac{1}{2} \times 3\frac{1}{2}$ _____

15) $5 \times 3\frac{1}{4}$ _____

16) $3\frac{1}{6} \times 7$ _____

17) $4\frac{2}{5} \times 6\frac{1}{8}$ _____

18) $3\frac{1}{3} \times 5\frac{4}{5}$ _____

19) $1\frac{1}{2} \times 6$ _____

20) $3\frac{4}{5} \times 7\frac{1}{2}$ _____

Division

Dividing fractions is the same as trying to find out what larger number your fraction is part of. For example, what is $\frac{3}{4}$ half of? This is a division problem. To understand what $\frac{3}{4} \div \frac{1}{2}$ really is, let's look at the following diagram (figure 4–8):

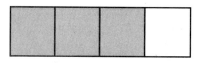

Figure 4–8

Three-fourths of the diagram is shaded in (indicated by the first three boxes). Then, to find out what larger number this is half of, we must double the picture to create two sets of three-fourths. Then we will know that our original set of three-fourths was only half of that (one set out of two sets drawn) (see figure 4–9).

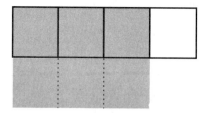

Figure 4–9

The larger number that $\frac{3}{4}$ is half of is six-fourths. Six shaded areas out of four possible, $\frac{6}{4}$.

To better understand division of fractions, you must realize that you are "undoing" multiplication (just as division is the opposite of multiplication in whole numbers). When we needed to know what larger number $\frac{3}{4}$ was half of, we had to double $\frac{3}{4}$. The opposite of finding a half is to double.

To divide fractions, you must first change all mixed numbers to improper fractions (as in multiplication). Then, invert the second fraction... that is, turn it upside down ($\frac{1}{2}$ becomes $\frac{2}{1}$). Be sure to always invert the **second** fraction, not the first one. Once this has been done, you can multiply the two fractions together.

Let's consider our example of $\frac{3}{4} \div \frac{1}{2}$.

$$\frac{3}{4} \div \frac{1}{2} = \qquad \text{becomes} \qquad \frac{3}{4} \times \frac{2}{1} = \frac{6}{4}$$

This answer is proof of our picture above and is the correct answer. It must now be reduced to lowest terms (that is, changed to a mixed number and then reduced): $\frac{6}{4} = 1\frac{2}{4} = 1\frac{1}{2}$

Example 1

$1\frac{1}{2} \div 1\frac{1}{5} =$

$1\frac{1}{2} \div 1\frac{1}{5} =$ First, change the mixed numbers to improper fractions.

$\frac{3}{2} \div \frac{6}{5} =$ Now invert (turn over) the second fraction and multiply (reduce first, if possible).

$\frac{\overset{1}{3}}{2} \times \frac{5}{\underset{2}{6}} = \frac{5}{4}$ $\begin{aligned} 1 \times 5 &= 5 \\ 2 \times 2 &= 4 \end{aligned}$

Reduce to lowest terms by changing back to a mixed number. $\frac{5}{4}$ becomes $1\frac{1}{4}$. The final answer is $1\frac{1}{4}$.
Remember: Once you have inverted the second fraction, the problem proceeds just like a fraction multiplication problem.

Problems   Divide the following pairs of fractions or mixed numbers and reduce to lowest terms.

1) $\frac{7}{8} \div \frac{3}{8}$ _____

2) $1\frac{2}{7} \div 1\frac{1}{2}$ _____

3) $2\frac{1}{2} \div 1\frac{2}{8}$ _____

4) $2\frac{1}{3} \div 1\frac{3}{4}$ _____

5) $1\frac{4}{11} \div 5$ (same as $\frac{5}{1}$) _____

6) $\frac{6}{7} \div 7$ _____

7) $1\frac{1}{3} \div \frac{11}{16}$ _____

8) $\frac{9}{10} \div \frac{2}{5}$ _____

9) $\frac{7}{9} \div \frac{7}{9}$ _____

10) $5 \div \frac{3}{5}$ _____

11) $5\frac{1}{3} \div \frac{2}{3}$ _____

12) $1\frac{3}{5} \div \frac{7}{10}$ _____

13) $1\frac{5}{7} \div \frac{3}{7}$ _____

14) $3\frac{1}{3} \div \frac{5}{6}$ _____

15) $2\frac{3}{4} \div \frac{1}{4}$ _____

Mixed Operations

Problems   Complete the following problems, using all necessary fraction operations.

1) Sally Reese used $3\frac{1}{2}$ ounces of antiseptic solution to clean her dental chair after each patient. On a day when she has 14 patients, how many ounces did she use altogether?   _____

2) Carla Crouse kept inventory of the cleaning supplies at Hanover House Elderly Care Center. The supplies are kept in three different closets throughout the facility. In the first closet, she found $2\frac{1}{4}$ bottles of floor cleaner, 6 bottles of hand soap, $3\frac{1}{8}$ bottles of all-purpose cleaner, and $4\frac{1}{2}$ bottles of glass cleaner. In the second closet, she found $1\frac{1}{8}$ bottles of floor cleaner, $3\frac{1}{2}$ bottles of hand soap, $4\frac{1}{4}$ bottles of all-purpose cleaner, and 5 bottles of glass cleaner. In the third closet, she found $5\frac{1}{8}$ bottles of floor cleaner, $3\frac{1}{8}$ bottles of hand soap, $4\frac{1}{16}$ bottles of all-purpose cleaner, and $3\frac{1}{4}$ bottles of glass cleaner.

   a. How much floor cleaner does she have altogether?   _____

   b. How much hand soap does she have altogether?   _____

   c. How much all-purpose cleaner does she have altogether?   _____

   d. How much glass cleaner does she have altogether?   _____

3) After Carla takes inventory, she must do whatever ordering is needed. If her goal is to keep at least two dozen bottles (altogether) of each cleaner in stock, how much of each cleaner does she need (leave in fractional amounts even though she would order full bottles)?

   a. floor cleaner?   _____

   b. hand soap?   _____

   c. all-purpose cleaner?   _____

   d. glass cleaner?   _____

4) Karen Smith is nursing supervisor over 5-East. A nurse's aide reports to her that Mr. Holcomb in Room 561 has used half of his IV already and there are only $\frac{3}{4}$ liters left. How much did the bag originally contain (use division)?   _____

5) Yvonne Farington began a diet at a weight of $192\frac{1}{8}$ pounds. She has lost 24 pounds and 5 ounces (5 ounces out of 16 ounces in a pound—$24\frac{5}{16}$ pounds). What does she weigh now?   _____

6) George is a veterinarian and is remodeling his office. The equipment supply company is willing to take his old items as a trade-in, and they will take $\frac{1}{4}$ off of the price of the new items he chooses. His new items total $1512. How much will he actually pay after they take $\frac{1}{4}$ off?   _____

7) Kevin Harrigan had $55\frac{1}{4}$ boxes of admission forms in inventory a month ago. Now he has $27\frac{5}{8}$ boxes left. How many boxes have been used?   _____

8) Janice Wagner gets paid time and a half for each hour of overtime. This is the same as getting paid for the hours overtime that she works plus half as many more (or $1\frac{1}{2}$ times the overtime hours). How many overtime hours will she get credit for working if she works $5\frac{1}{4}$ hours overtime?   _____

9) Nora Wilson has 3 uniforms for her new job. When she interviewed for the job, they said that she must have 5 different uniforms that she could wear. What fraction does she have? _____

10) Bob Nolton has $\frac{1}{2}$ of his required college credits to be a psychologist. He has earned $74\frac{1}{2}$ semester hours so far. How many does he need altogether? _____

Figure 4–10 Reprinted with permission from Simmers, *Diversified Health Occupations*, copyright 1983, Delmar Publishers Inc., Albany, NY

Chapter 5

Combined Operations

It is necessary at times to convert a number to a different form to make it more meaningful or useful in mathematical operations. The various forms covered in this section—decimal, fraction, and percent—are all interchangeable. We will learn how to convert correctly from one form to another to obtain an equality that can be further used to interpret or solve a problem.

Defining a Percent

The word **percent** means "per one hundred" or "parts of each hundred." The percent symbol after 23% means that 23 parts out of 100 possible are represented (figure 5–1).

Figure 5–1

You might say you've spent 25% more than you planned or you received a 5% raise. This is a comparison of the part (amount spent or amount of increase) to the whole amount (amount budgeted or amount currently earning).

Often standards are set on a 100-point scale. For example, grading is often figured based on 100 points. Sales, discounts, and statistics are often based on the percentage system as well (figure 5–2).

Percent to Decimal

Changing a percent to a decimal is a simple process when you understand what a percent is. Remember that a percent is defined as a certain number of parts out of 100 possible. That means that a percent is the same as a hundredth. In our number system, the hundredths place is two to the right of the decimal. Let's look at an example: 42% is 42 out of 100, or 42 hundredths. Forty-two hundredths is written as .42 (ending in the hundredths place—two to the right of the decimal). Therefore, 42% = .42

Figure 5–2

A standard rule for changing a percent to a decimal is to move the decimal two places to the left. This is also the same as the result you would get if you divide the number in the percent expression by 100. $42 \div 100 = .42$ (moved two places to the left).

Example 1

Change 12.5% to a decimal.

Move the decimal two places to the left and remove the percent sign.

$$12.5\% = .125$$

Example 2

Change 7% to a decimal.

Move the decimal two places to the left and remove the percent sign.

Remember: In a whole number, the decimal is understood to be at the right of the number.

$$7.\% = .07$$

Example 3

Change 125% to a decimal.

Move the decimal two places to the left and remove the percent sign.

$$125\% = 1.25$$

Problems   Change the following percents to decimals.

1) 55% _____	11) 125.5% _____	21) 71% _____
2) 7% _____	12) 11% _____	22) 200% _____
3) 65% _____	13) 10% _____	23) 5% _____
4) 60% _____	14) 4.25% _____	24) 68.7% _____
5) 100% _____	15) 8.1% _____	25) 45% _____
6) 78.33% _____	16) 873% _____	26) 13% _____
7) 49% _____	17) 372.2% _____	27) 1% _____
8) .05% _____	18) 35.1% _____	28) 99% _____
9) 74% _____	19) .361% _____	29) 3.6% _____
10) 13.5% _____	20) 15% _____	30) 68.5% _____

Decimal to Percent

Changing a decimal to a percent is the reverse process of changing a percent to a decimal. This can be accomplished by moving the decimal two places to the right. It is necessary to write a % sign after the number once the decimal placement has been changed.

Example 1

Change .475 to a percent.

Move the decimal two places to the right and add a % sign.

$$.475 = 47.5\%$$

Example 2

Change 3 to a percent.

Move the decimal two places to the right and add a % sign.

$$3. = 300\%$$

Example 3

Change .4 to a percent.

Move the decimal two places to the right and add a % sign.

$$.4 = 40\%$$

Problems   Change the following decimals to a percent.

1) .587	_____	11) .007	_____	21) 64	_____
2) 1.6	_____	12) .26	_____	22) .36	_____
3) .7314	_____	13) 5.3	_____	23) .0006	_____
4) .71	_____	14) .371	_____	24) .70	_____
5) 4	_____	15) 7.22	_____	25) .001	_____
6) .722	_____	16) 10	_____	26) .03	_____
7) .111	_____	17) 3.33	_____	27) .005	_____
8) 1	_____	18) .4	_____	28) .732	_____
9) .6	_____	19) .88	_____	29) 13.5	_____
10) .1	_____	20) .14	_____	30) 1.1	_____

Decimal to Fraction

To change a decimal to a fraction, you must determine what place value the decimal is holding. For example, .335 is holding the thousandths place (three places to the right of the decimal). Once you know its place value, the fraction is written with that place value as the denominator. In the case of .335, your fraction becomes $\frac{335}{1000}$ (335 parts out of 1000 possible). Then, this must be reduced, if possible.

$$.335 = \frac{335}{1000} \div \frac{5}{5} = \frac{67}{200}$$ The final answer is $\frac{67}{200}$.

Occasionally, the decimal will also have a whole number to the left of the decimal point. When this occurs, leave that whole number out of the changing process and then write it back in when you have reduced the fraction. For example, in 1.1 you have the whole number 1 and 1 tenth. The decimal part (.1) becomes $\frac{1}{10}$. Since this is lowest terms, rejoin the whole number 1 with the fraction for a final answer of $1\frac{1}{10}$.

Example 1

Change .885 to a fraction.

Determine the place value being held and rewrite as a fraction.

$$.885 = \frac{885}{1000} \div \frac{5}{5} = \frac{177}{200}$$

The final answer is $\frac{177}{200}$.

Example 2

Change 1.05 to a fraction.

1. Determine the place value being held and rewrite as a fraction (the decimal part only).

$$.05 = \frac{5}{100} \div \frac{5}{5} = \frac{1}{20}$$

2. Rejoin the whole number with the fraction. The final answer is $1\frac{1}{20}$.

Problems   Change the following decimals to fractions.

1) 4.56	_____	11) 3.7	_____	21) .007	_____
2) .8	_____	12) .015	_____	22) 1.6	_____
3) 25.1	_____	13) 7.50	_____	23) .42	_____
4) .62	_____	14) 71.85	_____	24) .71	_____
5) .154	_____	15) .26	_____	25) .88	_____
6) 1.68	_____	16) .25	_____	26) .027	_____
7) .080	_____	17) .16	_____	27) .22	_____
8) .789	_____	18) .35	_____	28) .64	_____
9) .3	_____	19) .04	_____	29) .025	_____
10) 3.5	_____	20) .75	_____	30) .002	_____

Fraction to Decimal

Changing a fraction to a decimal requires division. The numerator of the fraction becomes the dividend, and the denominator of the fraction becomes the divisor. Some fractions will divide out evenly (meaning that division will have no remainder). These decimals are called **terminating** decimals because they eventually end. Other fractions will not divide out evenly. These decimals are called **repeating** decimals because they will eventually develop a pattern that repeats itself in the decimal. Some repeating decimals tend to repeat very early in the division, and some do not. But in either case, since division will never come out evenly, there is a notation that can be used to express the answer obtained from division. When a number or group of numbers in your solution repeats, place a line over the number or group to indicate the repetition.

When working with a repeating decimal that does not repeat early in the division, it is sometimes more practical to round that number to a certain place value than it would be to complete the division. In this book, if you have not found a repeating pattern or a termination by the 100,000ths place (five places to the right of the decimal), round the answer to the nearest 10,000ths place (four places to the right of the decimal).

When a fraction is a mixed number, leave the whole number out of the changing process, then write it back in when you have solved the division and are writing the final answer.

Example 1

Change $\frac{1}{3}$ to a decimal.

$$
\begin{array}{r}
.333 \\
3\overline{)1.000} \\
-9 \\
\hline
10 \\
-9 \\
\hline
10 \\
-9 \\
\hline
1
\end{array}
$$

Use as many zero place holders as necessary.

The remainder 1 will continue to show up no matter how long we divide; this is a repeating decimal.

Since the 3 in .333 is the only digit repeating, we need only write one 3 with a line over it:

Final answer: $\frac{1}{3} = .\overline{3}$

Example 2

Change $\frac{3}{5}$ to a decimal.

$$
\begin{array}{r}
.6 \\
5\overline{)3.0} \\
-3.0 \\
\hline
0
\end{array}
$$

The remainder is zero; this is a terminating decimal.

Final answer: $\frac{3}{5} = .6$

Example 3

Change $\frac{1}{8}$ to a decimal.

$$
\begin{array}{r}
.125 \\
8\overline{)1.000} \\
-8 \\
\hline
20 \\
-16 \\
\hline
40 \\
-40 \\
\hline
0
\end{array}
$$

The remainder is zero; this is a terminating decimal.

Final answer: $\frac{1}{8} = .125$

Example 4

Change $\frac{1}{22}$ to a decimal.

```
       .04545
22)1.00000
   - 88
    ───
    120
   -110
    ───
    100
    - 88
     ───
    120
   -110
    ───
     10
```

The remainder 10 will continue to show up no matter how long we divide; this is a repeating decimal. Since the number group 45 is the repeating part of the decimal, we will put a line over both digits.

Final answer: $\frac{1}{22} = .0\overline{45}$

Problems   Change the following fractions to decimals. If division does not terminate or repeat before the 100,000ths place, round the answers to the nearest 10,000ths place.

1) $\frac{4}{5}$ _____ 6) $\frac{1}{16}$ _____ 11) $\frac{7}{8}$ _____

2) $\frac{5}{6}$ _____ 7) $\frac{7}{9}$ _____ 12) $\frac{3}{4}$ _____

3) $\frac{1}{10}$ _____ 8) $\frac{4}{8}$ _____ 13) $1\frac{1}{2}$ _____

4) $1\frac{5}{12}$ _____ 9) $3\frac{1}{5}$ _____ 14) $6\frac{1}{7}$ _____

5) $4\frac{3}{10}$ _____ 10) $5\frac{14}{15}$ _____ 15) $\frac{19}{20}$ _____

Percent to Fraction

There are two ways that a percent can be changed to a fraction. The first is to realize that a percent means a certain number of parts out of 100 possible. So, 15% means 15 out of 100, or $\frac{15}{100}$. This needs only to be reduced ($15\% = \frac{15}{100} = \frac{3}{20}$). This is, by far, the easiest way of changing a percent to a fraction and should be the first method that you try.

The second method is effective when the percent contains a decimal, like 14.5%. In this case, it is best to change the percent to a decimal first (as learned in a previous section) and then change that decimal to a fraction. So, to change 14.5% to a decimal, you must move the decimal point two places to the left and eliminate the % sign (14.5% = .145). Next, realize that .145 is in the 1000ths place, and rewrite the decimal as a fraction ($.145 = \frac{145}{1000}$). This must then be reduced ($.145 = \frac{145}{1000} = \frac{29}{200}$). The second method involves a few more steps, but it is the safest way to ensure a correct answer when the original percent is written in a complicated way.

Example 1

Change 100% to a fraction.

100% means 100 parts out of 100 possible.

$$100\% = \frac{100}{100} = 1$$

Example 2

Change 5% to a fraction.

5% means 5 parts out of 100 possible.

$$5\% = \frac{5}{100} = \frac{1}{20}$$

Example 3

Change 16.64% to a fraction.

Since this percent contains a decimal, use the second method (percent to decimal, then decimal to fraction). $16.64\% = .1664$

This is in the 10,000ths place:

$$.1664 = \frac{1,664}{10,000} = \frac{104}{625} \quad \text{(divide numerator and denominator by 16)}$$

Example 4

Change 450% to a fraction.

450% means 450 parts out of 100 possible.

$$450\% = \frac{450}{100} = \frac{9}{2} = 4\frac{1}{2} \quad \text{This must be written as a mixed number in lowest terms.}$$

Problems Change the following percents to fractions.

1) 50% _____	11) 42.5% _____	21) 6% _____
2) 102% _____	12) 96% _____	22) 10% _____
3) 11.5% _____	13) 14% _____	23) 7.25% _____
4) 33% _____	14) 70% _____	24) 110% _____
5) 4% _____	15) 65% _____	25) 12% _____
6) 47.2% _____	16) 51.25% _____	26) 11% _____
7) 8% _____	17) 52% _____	27) .25% _____
8) 1.6% _____	18) 1% _____	28) 32% _____
9) 116% _____	19) 26.75% _____	29) 38% _____
10) 3% _____	20) 162% _____	30) 18% _____

Fraction to Percent

To change a fraction to a percent, it is first necessary to change the fraction to a decimal by division (as we learned in a previous section). Then, change that decimal to a percent by moving the decimal two places to the right and adding a % sign.

Example 1

Change $\frac{3}{5}$ to a percent to determine the score on this quiz (see figure 5–3).

1. Change $\frac{3}{5}$ to a decimal by division.

$$
\begin{array}{r}
.6 \\
5\overline{)3.0} \\
-3.0 \\
\hline
0
\end{array}
$$

2. Change that decimal to a percent by moving the decimal two places to the right. $\frac{3}{5} = .6 = 60\%$ (add a zero place holder).

Example 2

Change $1\frac{7}{8}$ to a percent.

1. Change $\frac{7}{8}$ to a decimal by division.

$$
\begin{array}{r}
.875 \\
8\overline{)7.000} \\
-64 \\
\hline
60 \\
-56 \\
\hline
40 \\
-40 \\
\hline
0
\end{array}
$$

Remember the whole number 1, which has been set aside for now.

2. Change that decimal to a percent by moving the decimal two places to the right. $1\frac{7}{8} = 1.875 = 187.5\%$.

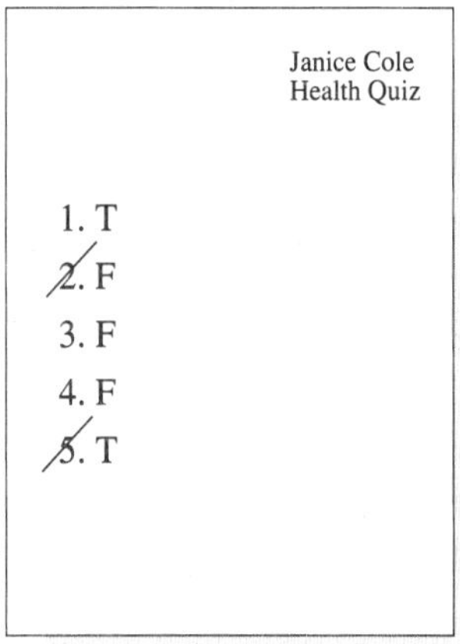

Janice Cole
Health Quiz

1. T
2. F
3. F
4. F
5. T

Figure 5–3   3 correct out of 5. $\frac{3}{5}$ = .6 and .6 = 60%.

Example 3

Change $\frac{1}{3}$ to a percent.

1. Change $\frac{1}{3}$ to a decimal by division.

$$\begin{array}{r} .33\dots \\ 3\overline{)1.00} \\ \underline{-9} \\ 10 \\ \underline{-9} \\ 1 \end{array}$$

This is a repeating decimal; it will need to be rounded to a certain place value.

2. Change that decimal to a percent by moving the decimal two places to the right.

$\frac{1}{3} = .\overline{3}$ Write this as a five-digit decimal for all problems in this book, then round to a four-digit number.

$\frac{1}{3} = .33333$ rounds to $.3333 = 33.33\%$

Problems   Change the following fractions to percents.

1) $\frac{4}{7}$	_____	6) $\frac{3}{8}$	_____	11) $1\frac{3}{4}$	_____
2 $\frac{5}{12}$	_____	7) $1\frac{4}{5}$	_____	12) $\frac{11}{12}$	_____
3) $\frac{1}{2}$	_____	8) $\frac{2}{5}$	_____	13) $\frac{1}{8}$	_____
4) $\frac{2}{3}$	_____	9) $\frac{1}{4}$	_____	14) $\frac{6}{10}$	_____
5) $\frac{1}{9}$	_____	10) $1\frac{1}{10}$	_____	15) $\frac{13}{14}$	_____

Percent-Base

To find a certain part of a whole or to find what percent of the whole is being represented or to find out what the whole number is to which you are comparing, you can work a percent-base problem. The percent represents the part you are comparing and the base is the whole to which you are comparing.

In a percent-base problem, there are three parts—the percent, the base, and the part (or percentage).

16	×	.25	=	4
16 is the base (whole)		.25 is the percent (changed to a decimal)		4 is the part (percentage)

This problem indicates that 25% (.25) of 16 is 4. Multiplication is used in this problem because of the word "of," and an equal sign replaces the word "is." Setting up the problem with the multiplication and equal signs is an important part of being able to solve it.

It will be necessary to solve a percent-base problem for each of the three possible answers. You might be asked: What is 15% of 200? Forty is 20% of what number? What percent of 80 is 24? Noticing the words "of" and "is" will help in setting up the problem and solving it. The word "what" indicates the missing answer. If the unknown answer is **not** alone on the left or right side of the equal sign, you must divide to find the solution. If the unknown answer **is** alone on the left or right side of the equal sign, you must multiply to find the solution.

Example 1

What is 15% of 200?

1. Put the problem into symbols and numbers.

   "what" = 15% × 200

2. Change the percent to decimal and solve.

   "what" = .15 × 200

3. Decide whether to multiply or divide (multiply because the unknown part is alone).

$$\begin{array}{r} 200 \\ \times\,.15 \\ \hline 1000 \\ 200 \\ \hline 30.00 \end{array}$$ 30 is 15% of 200

Example 2

Forty is 20% of what number?

1. Put the problem into symbols and numbers.

   40 = 20% × "what"

2. Change the percent to decimal and solve.

   40 = .20 × "what"

3. Decide whether to multiply or divide (divide because the unknown part is not alone).

$$\begin{array}{r} 200. \\ .20\overline{)40.00} \\ -40 \\ \hline 00 \end{array}$$ 40 is 20% of 200

Example 3

What percent of 80 is 24?

1. Put the problem into symbols and numbers.

   "what %" × 80 = 24

2. Change the percent to decimal and solve (we are looking for the percent… we can't change it now, but we must once we find it).

3. Decide whether to multiply or divide (divide because the unknown part is not alone).

$$\begin{array}{r} .3 \\ 80\overline{)24.0} \\ -240 \\ \hline 0 \end{array}$$

4. Change .3 to a percent (.3 = 30%).

   30% of 80 is 24

Problems   Complete the following percent-base problems, finding the missing answer by multiplication or division.

1) What is 18% of 50?_____

2) 18 is 300% of what number? _____

3) 50% of 16 is what? _____

4) 54 is 90% of what number? _____

5) What percent of 70 is 28? _____

6) What is 80% of 20?_____

7) What is 24% of 40?_____

8) What percent of 220 is 11? _____

9) 45 is 30% of what number? _____

10) 93 is what percent of 93? _____

11) What percent of 40 is 14? _____

12) What is 35% of 200?_____

13) What percent of 90 is 22.5? _____

14) What is 88% of 40?_____

15) 63 is 25% of what number? _____

16) 85% of 400 is what? _____

17) What percent of 100 is 47? _____

18) What is 18% of 250?_____

19) What is 55% of 220?_____

20) What is 325% of 44?_____

Mixed Operations

Problems   Fill in the blanks by converting among percents, decimals, and fractions.

	Fraction	Decimal	Percent
1)	$\frac{3}{4}$		
2)		.87	
3)			5.25%
4)		.005	
5)			16.5%
6)	$1\frac{7}{8}$		
7)	$\frac{3}{50}$		
8)		.124	
9)			312%
10)	$\frac{19}{20}$		
11)		4.5	
12)			2%
13)	$4\frac{1}{20}$		
14)		.0625	
15)			.04%
16)		.5	
17)	$\frac{5}{16}$		
18)			11.25%
19)		.006	
20)			5%

Chapter 6

Ratios and Proportions

Defining a Ratio

A **ratio** is an expression of comparison between two numbers. A ratio is usually written as one number before another with a colon in between. A ratio may also be expressed as a fraction, a decimal, or a quotient (one number divided by another).

The relationship between 1 and 4 can be written as 1:4 (standard notation), $\frac{1}{4}$ (fraction), .25 (decimal found by 1 ÷ 4), or 1 ÷ 4 (as a quotient).

Problems   Express the following ratios in the different forms requested. Reduce to lowest terms when possible.

	Ratio	Fraction	Decimal
1)	1:3	_____	_____
2)	2:5	_____	_____
3)	10:25	_____	_____
4)	2:15	_____	_____
5)	4:12	_____	_____
6)	1:8	_____	_____
7)	15:20	_____	_____
8)	85:100	_____	_____
9)	7:10	_____	_____
10)	5:9	_____	_____

Express the following ratios in these forms: standard, fraction, and quotient (Hint: Change to fraction in lowest terms first).

11)	.46	_____
12)	.15	_____
13)	.77	_____
14)	.05	_____
15)	1.35	_____

Defining a Proportion

A **proportion** expresses the relationship between two ratios. It is written as two ratios with an equal sign between. For example, 3:4 = 1:5. The four numbers in the proportion have special names.

The two outer numbers, in this case 3 and 5, are called the **extremes**. The two inner numbers, 4 and 1, are called the **means**. In a **true** (equal) proportion, the product of the means should equal the product of the extremes.

$$3:4 \ = \ 1:5$$

Means   Extremes

$4 \times 1 = 4$ $3 \times 5 = 15$ 4 does not equal 15; thus, this is not a true proportion.

Problems   Determine whether the following proportions are true. Show your work.

1) $4:5 = 12:15$ _____   11) $6:7 = 7:6$ _____

2) $1:5 = 3:14$ _____   12) $8:4 = 2:1$ _____

3) $16:3 = 4:.75$ _____   13) $12:8 = 3:2$ _____

4) $56:70 = 84:105$ _____   14) $5:7 = 10:13$ _____

5) $12:14 = 7:6$ _____   15) $20:30 = 2:3$ _____

6) $5:8 = 15:24$ _____   16) $14:16 = 8:7$ _____

7) $4:7 = 6:9$ _____   17) $10:18 = 5:9$ _____

8) $16:12 = 4:3$ _____   18) $5:10 = 4:8$ _____

9) $3:9 = 14:42$ _____   19) $6:9 = 8:18$ _____

10) $5:2 = 4:1$ _____   20) $16:15 = 4:5$ _____

Solving Proportions

If you know that a proportion is true (that the ratios are equal), you can solve for a missing part. This is very useful in the health field because it helps you solve problems based on another problem as your model. For example, if you know that a patient should get 1 ounce of medicine for every 20 pounds of body weight, you can use a proportion to determine how much to give a 180-pound man. For example, you can compare ounces to pounds in two ratios and set them equal.

$$\frac{oz}{lb} \quad \frac{1\ oz}{20\ lb} \ = \ \frac{no.\ of\ oz\ to\ give}{180\ lb}$$

Since this is a true proportion, the product of the means equals the product of the extremes.

Means = Extremes

$20 \times oz \ = \ 1 \times 180$

$20 \times oz \ = \ 180$ Divide both sides by 20. Dividing the left side by 20 results in 1 oz $(20 \div 20 = 1)$. Dividing the right side by 20 results in 9.

$oz \ = \ 180 \div 20$

$oz \ = \ 9$ The 180-lb man needs 9 oz of medicine.

Proportions do not always contain just whole numbers as the means and extremes. Occasionally, they may contain fractions, decimals, or more complicated expressions. These too can be solved for the unknown amount.

Example 1

Halterol is administered by the following formula: $\frac{1}{2}$ tablet per pint of blood lost. Josie recently lost three pints of blood. How much Halterol should be administered?

$$\frac{tablets}{pints} \qquad \frac{1/2}{1} = \frac{unknown\ "x"}{3}$$

Means = Extremes

$1\,x = \frac{1}{2}(3)$

$\qquad x = 1\frac{1}{2}$ Josie needs $1\frac{1}{2}$ tablets.

Example 2

Solve $5 : (x - 2) = 10:6$

Means = Extremes	
$10(x - 2) = 5 \times 6$	Multiply everything in the parentheses by 10.
$10x - 20 = 30$	Add 20 to both sides.
$10x - 20 + 20 = 30 + 20$	A negative 20 plus 20 equals 0.
$\dfrac{10x}{10} = \dfrac{50}{10}$	Divide both sides by 10.
$x = 5$	

It is wise to check your answers when you have solved a proportion. You should be able to replace the answer you found back into the original proportion, multiply the means and extremes, and the result will be a true proportion. If it is not, recheck your work.

Problems   Solve for x in the following proportions. Check your work.

1) $4:5 = 12:x$ _____

2) $8:x = 28:7$ _____

3) $4.5:5 = x:10$ _____

4) $8:(x - 3) = 12:6$ _____

5) $7:x = 21:24$ _____

6) $13:x = 52:4$ _____

7) $19:x = 38:40$ _____

8) $(x + 3):8 = 8:4$ _____

9) $7:10 = x:3.5$ _____

10) $13:3 = 13:x$ _____

11) $x:7 = 12:42$ _____

12) $9:10 = x:15$ _____

13) $24:30 = x:5$ _____

14) $2.5:10 = x:24$ _____

15) $6:9 = 10:x$ _____

16) $3:4 = x:16$ _____

17) $x:6 = 7:42$ _____

18) $6:15 = 3:x$ _____

19) $4:5 = x:15$ _____

20) $4:8 = 8:x$ _____

Chapter 7

Measurement

Conversion Facts

Volume, length, and mass are the three main types of measurement. Chapter 8 deals with the study of these measurements in the metric system, and this chapter will look at them in the English system. The English system is the system most commonly used in the United States. In the health field, it is still widely used to measure the height and weight of a patient and to determine food portions (see figures 7–1a, 7–1b, 7–1c).

The most common volume measurements in the English system include the gallon, half gallon, quart, pint, cup, ounce, tablespoon, and teaspoon. The most commonly used length measurements are the mile, yard, foot, and inch. The mass measurements include pounds and ounces.

To understand the English system and use it, it is important to know some basic facts about the relationship of one measurement to another. It is also helpful to have a standard set of abbreviations to work with.

Figure 7–1a Reprinted with permission from Caldwell, *Nursing Assistant*, 5E, copyright 1989, Delmar Publishers Inc., Albany, NY.

Figure 7–1b Reprinted with permission from Caldwell, *Nursing Assistant*, 5E, copyright 1989, Delmar Publishers Inc., Albany, NY.

Figure 7–1c Reprinted with permission from Caldwell, *Nursing Assistant*, 5E, copyright 1989, Delmar Publishers Inc., Albany, NY.

Volume

Abbreviations:	gallon = gal
	half gallon = $\frac{1}{2}$ gal
	quart = qt
	pint = pt
	cup = c
	ounce = oz
	tablespoon = T, tbl, tbsp
	teaspoon = t, tsp
Equivalent measures:	1 gallon = 2 half gallons = 4 quarts
	$\frac{1}{2}$ gallon = 2 quarts
	1 quart = 2 pints
	1 pint = 2 cups
	1 cup = 8 ounces
	1 ounce = 2 tablespoons
	1 tablespoon = 3 teaspoons

Length

Abbreviations:	mile = mi
	yard = yd
	foot = ft or '
	inch = in or "
Equivalent measures:	1 mile = 1760 yards or 5280 feet
	1 yard = 3 feet
	1 foot = 12 inches

Mass

Abbreviations:	pound = lb or #
	ounce = oz
Equivalent measures:	1 pound = 16 ounces

The English system of measurement is not like the metric system, which follows a pattern (multiples of ten). It has been developed more randomly (**three** teaspoons in a tablespoon; **four** quarts in a gallon; **16** ounces in a pound). Therefore, when we want to convert from one measurement to another, we cannot simply move the decimal (an advantage to multiplying or dividing by ten). Instead, we must multiply or divide by different numbers for each different conversion. These numbers are called **conversion numbers**.

To find the conversion number, we will first write all of the same type measurements (volume, mass, or length) in order from largest to smallest in a column.

Example

1 quart = _____ ounces

gallon	cup
half gallon	ounce
quart	tablespoon
pint	teaspoon

Then, we will write the facts that we know about these measurements along the side in an abbreviated way.

Example

1 quart = _____ ounces

*	gallon	(there is nothing above on list)
2	half gallon	(gallon from above = 2 half gal)
2	quart	(half gal from above = 2 qt)
2	pint	(quart from above = 2 pt)
2	cup	(pint from above = 2 c)
8	ounce	(cup from above = 8 oz)
2	tablespoon	(ounce from above = 2 tbl)
3	teaspoon	(tbl from above = 3 tsp)

Now we need to determine what two measurements concern us in the conversion problem we are doing. In our example, this is quart and ounce. So we must find these measurements on the list.

2	quart	(half gal from above = 2 qt)
2	pint	(quart from above = 2 pt)
2	cup	(pint from above = 2 c)
8	ounce	(cup from above = 8 oz)

It is important to ignore the top number in this new list because it is information about the half gallon from above. We need all of the numbers except the top one (2, 2, and 8). Multiply these numbers together ($2 \times 2 \times 8 = 32$). This is your conversion number.

Once you have found your conversion number, the following rules will help you to complete the conversion.

1. When converting from a large unit to a smaller one, multiply by the **conversion number**.

2. When converting from a small unit to a larger one, divide by the **conversion number**.

In our example, 1 quart = _____ ounces, we are converting from a larger unit to a smaller unit. (This is made obvious because we are going **down** the list that we wrote—from largest to smallest). Since we are going from large to small, we should multiply. **One** quart $\times$ **32** = 32 ounces. 1 quart = 32 ounces.

This system of abbreviation with numbers is just a device to help you remember the equivalent facts about measurement. Let's look again at the list with the numbers along the side.

Volume

*	gallon	(there is nothing above on list)
2	half gallon	(gallon from above = 2 half gal)
2	quart	(half gal from above = 2 qt)
2	pint	(quart from above = 2 pt)
2	cup	(pint from above = 2 c)
8	ounce	(cup from above = 8 oz)
2	tablespoon	(ounce from above = 2 tbl)
3	teaspoon	(tbl from above = 3 tsp)

There are seven numbers there in the complete list: 2, 2, 2, 2, 8, 2, and 3. Think of these seven numbers as a telephone number 222–2823. This is easy to remember and will help you convert the volume measurements. For length, the list is much shorter and not too difficult to remember. And for mass, you need remember only the number 16 (16 ounces in a pound).

Length

*	mile
1760	yard	(mile from above = 1760 yd)
3	foot	(yard from above = 3 ft)
12	inch	(foot from above = 12 in)

Mass

*	pound
16	ounce	(pound from above = 16 oz)

Example 1

3 c = _____ tbl

1. Write the volume list.

*	gallon
2	half gallon
2	quart
2	pint
2	cup
8	ounce
2	tablespoon
3	teaspoon

2. Find the two measurements on the list (c, tbl).

2	cup
8	ounce
2	tablespoon

3. Ignore the top number, multiply the rest ($8 \times 2 = 16$).

4. You are going from large to small, so multiply (**3 c × 16** = 48).

5. Solution: 3c = 48 tbl

Example 2

3520 ft = _____ mi

1. Write the length list.

*	mile
1760	yard
3	foot
12	inch

2. Find the two measurements on the list (ft, mi).

*	mile
1760	yard
3	foot

3. Ignore the top number, multiply the rest. (The top number is the *... ignore it. $1760 \times 3 = 5280$)

4. You are going from small to large, so divide. You have the choice here of: $\frac{3520}{5280}$ fraction to be reduced or division.

    $$3520 \text{ ft} \div 5280 = \tfrac{2}{3}$$

5. Solution: $3520 \text{ ft} = \tfrac{2}{3}$ mi

Example 3

How many ounces are in a gallon?
In this example, it is necessary to first restate the problem in the form, "what we're given... what we're trying to find."

1. Rewrite as: 1 gallon = _____ ounces

2. Write the volume list.
    - * gallon
    - 2 half gallon
    - 2 quart
    - 2 pint
    - 2 cup
    - 8 ounce
    - 2 tablespoon
    - 3 teaspoon

3. Find the two measurements on the list (gal, oz).
    - * gallon
    - 2 half gallon
    - 2 quart
    - 2 pint
    - 2 cup
    - 8 ounce

4. Ignore the top number, multiply the rest (ignore the *... $2 \times 2 \times 2 \times 2 \times 8 = 128$).

5. You are going from large to small, so multiply: **1** gal $\times$ **128** = 128

6. Solution: 1 gal = 128 oz

Volume Measurement

In converting volume measurements, you will always need the list of volume measurements in order from largest to smallest.

- * gallon
- 2 half gallon
- 2 quart
- 2 pint
- 2 cup
- 8 ounce
- 2 tablespoon
- 3 teaspoon

Chapter 8

Metric System

Explaining the Metric System

The metric system is a system of measurement based on the number 10. There are 10 smaller units in each larger unit. The metric system has three basic units of measure:

the gram   (measures mass or weight)

the liter   (measures volume or liquid)

the meter (measures length or distance)

Each of the units can be multiplied by 10 to produce increasingly larger units or divided by 10 to produce smaller units.

Let's look at the meter. It is $39\frac{3}{8}$ inches long. It is similar in size to a yardstick. It can be used to measure the length of a car or the height of an adult, but it is not useful for measuring the size of a sheet of paper or a paperclip. We need a smaller unit of measure. If we divide the meter by 10, it produces a unit called a decimeter. This can be used to measure smaller items. For very small items, we may measure in millimeters (which is a meter divided by 1,000... or a meter divided by 10 three times). For very large items or distances, we may measure in kilometers (which is a meter multiplied by 1,000... or multiplied by 10 three times).

The various length measurements are:

kilometer   hectometer   dekameter   meter   decimeter   centimeter   millimeter

Larger .... Smaller

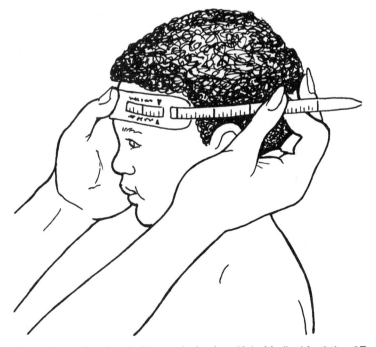

Figure 8–1   Reprinted with permission from Keir, *Medical Assisting*, 2E, copyright 1989, Delmar Publishers Inc., Albany, NY.

The various volume measurements are:

kiloliter   hectoliter   dekaliter   liter   deciliter   centiliter   milliliter

Larger ....Smaller

The various mass measurements are:

kilogram   hectogram   dekagram   gram   decigram   centigram   milligram

Larger ....Smaller

The three types of measurement are interrelated in the metric system. For example, if you made a cube that measured one centi**meter** and filled it with water, the amount it would hold is one milli**liter**. Then if you weighed that cube of water, it would weigh one **gram** (see figure 8–2).

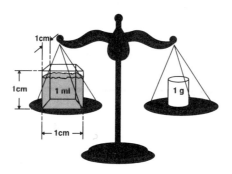

Figure 8–2   1 cm³ = 1 ml of water = 1 g.

Prefixes and Abbreviations

The gram, meter, and liter are considered the base for each measurement. Any larger or smaller measurement is shown by using a prefix in front of these words (as shown above). Let's look more closely at the prefixes:

- kilo-   base × 1,000
- hecto-   base × 100
- deka-   base × 10

- deci-   base / 10
- centi-   base / 100
- milli-   base / 1,000

Base (meter, liter, or gram)

or, another way of looking at this list is:

- kilo-   base × 1,000
- hecto-   base × 100
- deka-   base × 10

- deci-   base × 0.1
- centi-   base × 0.01
- milli-   base × 0.001

Base (meter, liter, or gram)

To save time and standardize the use of metrics, abbreviations are often used. The abbreviations for the base units are:

meter = m   liter = l   gram = g

The abbreviations for the prefixes are:

kilo = k   deka = dk   centi = c
hecto = h   deci = d   milli = m

A prefix can never be used alone. It must have a base unit with it to indicate whether you are measuring length, volume, or mass.

Therefore, the abbreviations for the prefixes above must be combined with an abbreviation for a base unit as follows:

Length	Abbreviation
kilometer	km
hectometer	hm
dekameter	dkm
meter	m
decimeter	dm
centimeter	cm
millimeter	mm

Notice that m alone always means meter, since milli-cannot stand alone.

Figure 8–3   Metric length

Figure 8–4   Metric volume

Volume	Abbreviation
kiloliter	kl
hectoliter	hl
dekaliter	dkl
liter	l
deciliter	dl
centiliter	cl
milliliter	ml

Mass	Abbreviation
kilogram	kg
hectogram	hg
dekagram	dkg
gram	g
decigram	dg
centigram	cg
milligram	mg

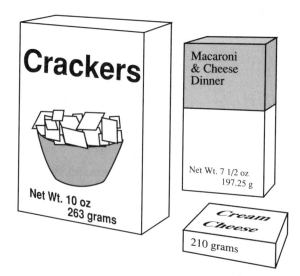

Figure 8–5   Metric mass

Problems

1) Fill in the abbreviations for the following.

   a. kilogram   _____   f. centigram   _____

   b. deciliter   _____   g. hectoliter   _____

   c. dekameter   _____   h. milligram   _____

   d. liter   _____   i. gram   _____

   e. millimeter   _____   j. dekaliter   _____

2) Match the base unit with what it measures by drawing a connecting line.

   meter   volume

   liter   mass

   gram   length

3) Using abbreviations, fill in the appropriate measurement for each of these descriptions.

   a. 1,000 grams   _____   f. 10 meters   _____

   b. 100 liters   _____   g. $\frac{1}{100}$ liter   _____

   c. $\frac{1}{10}$ meter   _____   h. 1,000 meters   _____

   d. $\frac{1}{1,000}$ liter   _____   i. 10 grams   _____

   e. 1 gram   _____   j. 1 liter   _____

Conversions in the Metric System

It is important in working with metric measurements to be able to convert from one measure to another. For example, how many centimeters are in a hectometer? Or how many milliliters are in a liter?

Because all of the measurements are related by the number ten, we can convert measurements by moving the decimal.

Example 1

Let's find out how many centimeters are in a hectometer. This can be rewritten as: 1 hm = _____ cm. This takes us from what we are given (1 hm) to what we are trying to find (how many cm). We need a list of measurements for length (written largest to smallest):

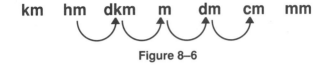

Figure 8–6

To convert from hm to cm, we are going four units to the right. This means we should move our decimal four places to the right. Remember that in a whole number, the decimal begins at the right of the number.

Figure 8–7

So, 1 hm = 10,000 cm

Example 2

How many milliliters are in 6.9 liters?

1. First restate the problem as 6.9 l = _____ ml

2. Write down the list of measurements for volume.

    kl   hl   dkl   l   dl   cl   ml

3. Determine the number of units you are moving and in which direction.

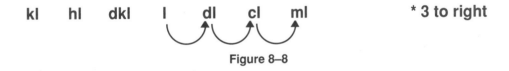

Figure 8–8

4. Move the decimal that number of places in the same direction.

Figure 8–9

So, 6.9 l = 6,900 ml

Example 3

How many hg are in 423 dg?

1. Restate the problem (notice that the words "how many" point us to what we are trying to find).

    423 dg = _____ hg

2. Write down the list of measurements for mass.

    kg   hg   dkg   g   dg   cg   mg

3. Determine the number of units you are moving and in which direction.

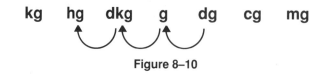

Figure 8–10

4. Move the decimal that number of places in the same direction.

423. ———————————— **.423**

Figure 8–11

So, 423 dg = .423 hg

Volume Measurement

In converting volume measurements, you will always need the list of volume measurements.

kl   hl   dkl   l   dl   cl   ml

1) 29 dkl = _____ dl   9) 2,947 ml = _____ cl

2) 18.6 l = _____ ml   10) 24.7 dkl = _____ kl

3) 37,423 ml = _____ dl   11) 932.6 l = _____ hl

4) 56.2 kl = _____ hl   12) 46.5 dl = _____ cl

5) .14 l = _____ cl   13) .26 cl = _____ dkl

6) 83.92 cl = _____ hl   14) 4.553 l = _____ dl

7) 6 dkl = _____ l   15) 811.6 kl = _____ hl

8) 15 dl = _____ ml

In problems 16–20, be sure to restate the problem as shown in the examples.

16) How many dl are in 16 l?   _____

17) A hl equals how many l?   _____

18) How many dekaliters are in 276.8 deciliters?   _____

19) How many milliliters are in a 2-liter bottle?   _____

20) How many hl are in 27 dkl?   _____

Length Measurement

In converting length measurements, you will always need the list of length measurements.

km   hm   dkm   m   dm   cm   mm

1) 187.6 km = _____ dkm   9) 43 km = _____ m

2) 2965 cm = _____ m   10) 8.167 m = _____ cm

3) 42.5 hm = _____ km   11) 14,382 hm = _____ km

4) 19,653 cm = _____ dkm   12) 9 mm = _____ dm

5) 29 m = _____ mm   13) 7,676 dm = _____ hm

6) 14.37 hm = _____ dm   14) 2.96 km = _____ dkm

7) 49.8 cm = _____ hm   15) 1.119 cm = _____ mm

8) 2963 mm = _____ dm

In problems 16–20, be sure to restate the problem as shown in the examples.

16) How many m are in a km? _____

17) How many dekameters are in 6,921 centimeters? _____

18) 18 dkm equals how many mm? _____

19) A patient 17.5 dm tall is how many meters tall? _____

20) How many decimeters are in 2.8 hectometers? _____

Mass Measurement

In converting mass measurements, you will always need to list of mass measurements.

kg	hg	dkg	g	dg	cg	mg

1) 8.63 kg = _____ g   9) 863 g = _____ cg

2) 49,682 cg = _____ dkg   10) 2,111 hg = _____ dg

3) 425 dg = _____ dkg   11) 91.164 cg = _____ hg

4) 84.1 hg = _____ cg   12) 4 g = _____ kg

5) 436 kg = _____ mg   13) 56.11 dg = _____ dkg

6) 89,632 mg = _____ cg   14) 81 kg = _____ hg

7) 42.31 dkg = _____ kg   15) 534 g = _____ mg

8) 5.85 mg = _____ dg

In problems 16–20, be sure to restate the problems as shown in the examples.

16) How many cg are in 15 dkg? _____

17) How many grams are in 4.65 kilograms? _____

18) How many mg are in a gram? _____

19) How many dg are in 4 hg? _____

20) How many kilograms are in 1654 dekagrams? _____

Metric/English Conversions

Often in our use of the metric system, it is helpful to know the relationship of metric measurements to their English counterparts.

In the health fields, volume measurements are particularly important. You need to be able to convert the liquid intake and output of your patients into the metric system for recording on their charts.

To do this, it is necessary to know some conversion facts (these are the approximations most commonly used):

Volume

1 quart = 1 liter = 1,000 ml or cc   1 tbl = 15 ml or cc
1 pint = 500 ml or cc   1 tsp = 5 ml or cc
1 cup = 240 ml or cc   1 drop (1 gtt) = .0667 ml or cc
1 ounce = 30 ml or cc

Length

1 mile = 1601.6 m   1 foot = .31 m
1 yard = .91 m   1 inch = .025 m

Mass

1 pound = .454 kg   1 ounce = .028 kg

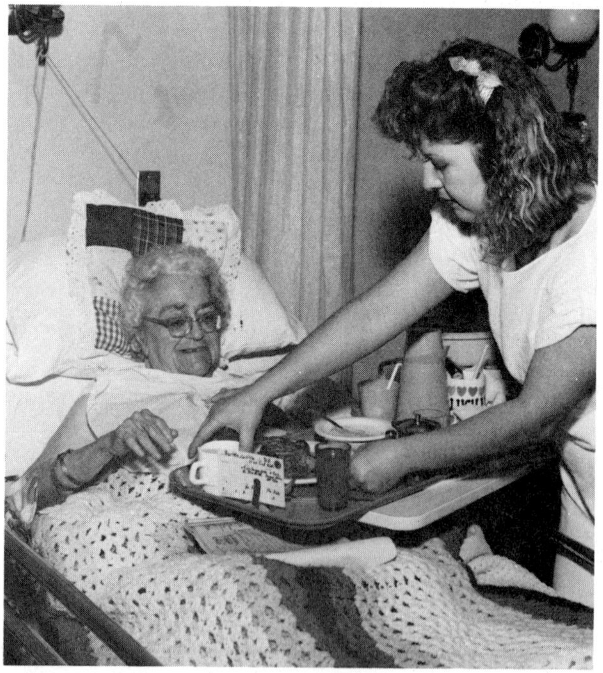

Figure 8–12   Reprinted with permission from Caldwell, *Nursing Assistant*, 5E, copyright 1989, Delmar Publishers Inc., Albany, NY.

One simple rule is also helpful. When changing English units to metric using the above conversion facts, multiply by the appropriate conversion number. When changing metric to English, divide.

English to metric = multiply
metric to English = divide

Example 1

How many ml are in 4 pints?

1. Restate the problem.

   4 pt = _____ ml

2. This is English to metric, so multiply.

3. Determine which conversion fact to use (1 pt = 500 ml).

4. 4 pt × 500 ml = 2000 ml

So, 4 pt = 2000 ml

Example 2

How many meters are in 100 yards?

1. Restate the problem.

   100 yd = _____ m

2. This is English to metric, so multiply.

3. Determine which conversion fact to use (1 yd = .91 m).

4. 100 yd × .91 m = 91

So, 100 yds = 91 m

Example 3

How many pounds are in 68.1 kg?

1. Restate the problem.

   68.1 kg = _____ lb

2. This is metric to English, so divide.

3. Determine which conversion fact to use (1 lb = .454 kg).

4. 68.1 kg / .454 kg = 150

So, 68.1 kg = 150 lb

Problems   Complete the following conversions.

1) a. 45 drops = _____ ml   e. 2.25 pt = _____ ml

   b. 3 oz = _____ ml   f. 3 tsp = _____ ml

   c. 1.5 c = _____ ml   g. 4 c = _____ ml

   d. 13 tbl = _____ ml   h. 10 gtt = _____ ml

2) a. 270 ml = _____ oz   4) a. 3.1 m = _____ ft

    b. 45 ml = _____ tsp   b. 75 m = _____ in.

    c. 3000 ml = _____ pt   c. 1.365 m = _____ yd

    d. 1920 ml = _____ c   d. 1.55 m = _____ ft

    e. 1 ml = _____ gtt   5) a. 6 lb = _____ kg

    f. 90 ml = _____ tbl   b. 150 lb = _____ kg

    g. 45 ml = _____ oz   c. 16 oz = _____ kg

    h   120 ml = _____ c   d. 4 oz = _____ kg

3) a. 6 yd = _____ m   6) a. 2.66 kg = _____ oz

    b. 20 in = _____ m   b. 2.8 kg = _____ oz

    c. 2 ft = _____ m   c. 27.24 kg = _____ lb

    d. 6 in = _____ m   d. 4.086 kg = _____ lb

In problems 7–12, round answers to the nearest tenth.

7) Amelia Armand was weighed at the doctor's office. She weighed 129
    pounds. What is her weight in kg?   _____

8) Steve Phillips is 70" tall. What is his height in meters?   _____

9) Kathy Dear drank 3 cups of water. How many cc of water did she drink?   _____

10) Jim Craig is moving 7 kg of supplies to storage. He needs to determine which cart to use.
    Should he select one which holds (show your work):

    a. 100 lb   _____

    b. 150 lb   _____

    c. 200 lb   _____

    d. 250 lb   _____

11) Abbey Hale used 650 ml of antiseptic solution to clean the dental chair.
    How many pints is this?   _____

12) Sharon Hogan ran in a 5 km run. How many miles is this? (Hint: Change
    5 km to meters first.)   _____

Chapter 9

Roman Numerals

Roman Numerals

The number system that we use daily and are familiar with is the system of **Arabic** numerals (0–9 or any combination of these digits). In medication, prescriptions, and other occasional uses, however, it is necessary to know **Roman** numerals. The Roman numeral system uses letters to represent numeric values (figure 9-1).

The basic numerals in the Roman system are:

Roman	Arabic
I	one
V	five
X	ten
L	fifty
C	one hundred
D	five hundred
M	one thousand

These numerals can be used in combination to represent all other values. To use Roman numerals, however, there are a few basic rules that must be followed.

Rule 1

When two Roman numerals of the same or decreasing value are written beside each other, the values are added together.

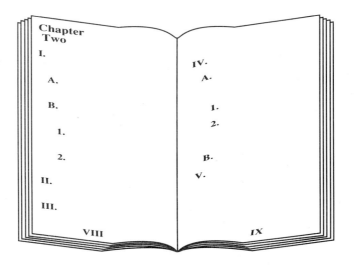

Figure 9–1   One common use of Roman numerals is in textbooks.

Example

VI = 5 + 1 = 6   XX = 10 + 10 = 20

Rule 2

When Roman numerals of increasing value are written beside each other, the smaller value is subtracted from the larger value.

Example

IV = (small before large, so subtract) 5 – 1 = 4
XC = 100 – 10 = 90

Rule 3

No more than three Roman numerals of the same value should be used in a row. If you need a larger value, it can be written using the subtraction principle (rule 2).

Example

To write the number 4, do not use IIII. Instead, use IV (5 – 1 = 4).
To write the number 59, do not use LVIIII. Instead, use L for 50, then IX (10 – 1) for 9. LIX = 59

Rule 4

A line written over a Roman numeral changes its value to 1,000 times its original value.

Example

$\overline{X}$ = 10,000

Table 9–1   Basic Numeric Values

Arabic Numerals	Roman Numerals	Arabic Numerals	Roman Numerals	Arabic Numerals	Roman Numerals
1	I	8	VIII	60	LX
2	II	9	IX	70	LXX
3	III	10	X	80	LXXX
4	IV	20	XX	90	XC
5	V	30	XXX	100	C
6	VI	40	XL	500	D
7	VII	50	L	1,000	M

Converting Roman to Arabic

To convert Roman numerals to Arabic, you must be able to separate the Roman numerals into manageable units first. The easiest way to do this is to make a distinction between the groups that need to be added together and the groups that require subtraction.

First, scan the Roman numerals, looking for any groups where a smaller value is written before a larger value. These indicate subtraction. Then, total the values from left to right.

Example 1

Convert XCIV to an Arabic number.

1. XC is a subtraction group (100 – 10 = 90). IV is also a subtraction group (5 – 1 = 4).

2. Now total the values from left to right: 94.

Example 2

Convert CDLXXVII to an Arabic number.

1. CD is a subtraction group (500 – 100 = 400).

2. Now total 400 + 50 + 10 + 10 + 5 + 1 + 1 = 477

Realize before you begin conversion that it is necessary to know the values of the various Roman numerals so that you can recognize subtraction groups. Failure to recognize a smaller value written before a larger one will result in an incorrect answer. In medications and other health uses, great care must be taken to obtain an accurate value.

Problems   Convert the following Roman numerals to Arabic numerals.

1)	DCCXLVII	_____	16)	MCMLXIV	_____
2)	CXLVIII	_____	17)	XXXIV	_____
3)	XCIV	_____	18)	VII	_____
4)	XXXIX	_____	19)	CV	_____
5)	XXVI	_____	20)	CCCLXXXIX	_____
6)	MCMLXXXIX	_____	21)	CMXCVI	_____
7)	CM	_____	22)	XIX	_____
8)	XXX	_____	23)	LXXIX	_____
9)	XIV	_____	24)	DCCL	_____
10)	DCCXLIV	_____	25)	MMDCCLIV	_____
11)	MCML	_____	26)	CL	_____
12)	CCCLIV	_____	27)	MCDLVII	_____
13)	XVIII	_____	28)	DCLIX	_____
14)	IL	_____	29)	XXIV	_____
15)	CDXX	_____	30)	$\overline{\text{LX}}$	_____

Converting Arabic to Roman

To convert Arabic numerals to Roman, you must be able to separate the Arabic numerals into manageable units first. The easiest way to do this is to make a distinction among the different place values in our number system (ones, tens, hundreds, etc.).

Going from left to right, write each place value using Roman numerals. Be sure to follow the four rules for writing Roman numerals correctly.

Example 1

Convert 1,768 to a Roman numeral.

1. Separate 1,768 into   1,000,   then write in Roman   M
                                  700   DCC
                                    60   LX
                                    8   VIII

2. Write out the Roman numeral as one answer (leave no spaces between the different parts). The final answer is: MDCCLXVIII

Example 2

Convert 479 to a Roman numeral.

1. Separate 479 into   400,   then write in Roman   CD
                             70   LXX
                             9   IX

2. Write out. The final answer is: CDLXXIX

In some cases, a shortcut can be made if you recognize that you are just one digit away from a major Roman value (D, C, L, etc.). For example, in 799, you are one digit away from an even hundred. You can do this as 700 (DCC) and 99 (100 – 1 or IC). The answer DCCIC is shorter than the answer obtained the traditional way:

                                   700   DCC (DCCICIX)
                                   90   IC
                                   9   IX

Problems   Convert the following Arabic numerals to Roman.

1) 522	_____	11) 1,415	_____	21) 599	_____
2) 133	_____	12) 609	_____	22) 14	_____
3) 1,099	_____	13) 457	_____	23) 813	_____
4) 999	_____	14) 36	_____	24) 575	_____
5) 82	_____	15) 79	_____	25) 163	_____
6) 10,673	_____	16) 376	_____	26) 26	_____
7) 10	_____	17) 77	_____	27) 155	_____
8) 510	_____	18) 268	_____	28) 945	_____
9) 5,378	_____	19) 362	_____	29) 7812	_____
10) 307	_____	20) 16	_____	30) 718	_____

Chapter 10

Medication Dosages

Abbreviations

Abbreviations are very common in prescriptions and orders from a physician. Abbreviations are used to indicate the frequency of the medication or other order, the means by which the medication is to be administered (orally, intravenously, etc.), the dosage to be given, and other necessary information. It is important that you become familiar with the abbreviations in table 10–1 (page 80).

Adult Oral Dosages

An **oral** medication is one taken by mouth. Examples of oral medication include syrups, tablets, capsules, lozenges, and powders. Oral administration is the most common way to give medication (see figures 10–1a and 1b).

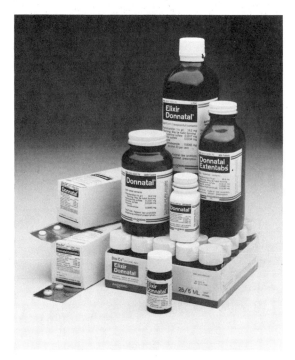
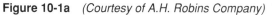

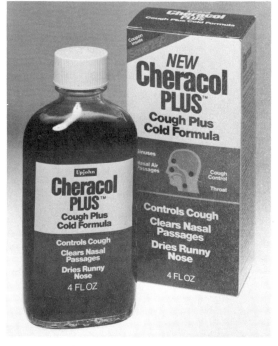

Figure 10-1a   *(Courtesy of A.H. Robins Company)*   **Figure 10-1b**   *(Courtesy of Upjohn Company)*

Table 10–1 Common Medical Abbreviations

Abbreviation	Meaning	Abbreviation	Meaning
aa	of each	o.u.	both eyes
a.c.	before meals	oz	ounce
ad lib	as desired	OTC	over the counter (drugs)
agit	shake, stir	p.c.	after meals
alt dieb	alternating days	PL	placebo
am	morning	pm	afternoon
aq	water	p.o.	by mouth
b.i.d.	two times a day, twice daily	pr	per rectum
c	with	PRN	as required or as necessary
cap	capsule	pulv	powder
cc	cubic centimeter	q.	every
chem	chemotherapy	q.d.	every day
dil	dilute	q.h.	every hour
disc.	discontinue	q.2h	every 2 hours
disp	dispense	q.3h	every 3 hours
fl	fluid	q.4h	every 4 hours
FM	flowmeter	q.i.d.	four times a day
garg	gargle	q.m.	every morning
gr.	grain	q.n.	every night
gtt(s)	drop(s)	q.o.d.	every other day
(H)	hypodermic	q.s.	qty. sufficient
h, hr	hour	R	rectal
h.s.	hour of sleep	Rx	"take thou"
H_2O	water	$\bar{s}$	without
IM	intramuscular	S, Sig	give the following instructions
IV	intravenously		
med	medicine	SC, subq	subcutaneous
mn	midnight	$\bar{\bar{ss}}$	one-half
MO	mineral oil	stat	immediately
MOM	milk of magnesia	t.i.d.	three times a day
MTD	maximum tolerated dose	tinct	tincture
noc, noct	night	TO	telephone order
NPO	nothing by mouth	tus	cough
N/S	normal saline	U	unit
O	pint	vag	vagina
O_2	oxygen	ves	bladder
OD	overdose	VO	verbal order
o.d.	right eye	W/O	water in oil

No drug should be administered except with physician's orders. An order is commonly called a **prescription**, and it gives the following information (see figure 10–2):

the name of the drug

the amount prescribed (dosage)

the frequency with which it should be administered

the route of administration

the purpose for which it is prescribed

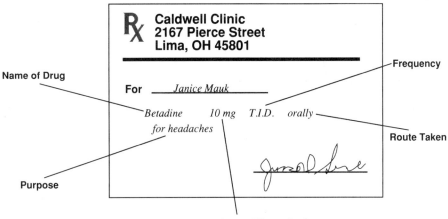

Figure 10–2

The amount prescribed by the physician is often written in grams and milligrams. You will need to be able to determine how many grams or milligrams are in the tablets prescribed, in the capsules prescribed, or in each ml of liquid prescribed.

The use of proportions is especially important in computing drug dosages. The proportion we will use throughout the chapter is this:

$$\frac{known\ unit\ on\ hand}{known\ dosage\ form} = \frac{dose\ ordered}{unknown\ amount\ to\ be\ given}$$

You should memorize this proportion and be able to use it in all dosage situations. Let's break it down a bit to better understand its use.

Known unit on hand: This is the amount of grams or milligrams that the selected drug contains in the known dosage form.

Known dosage form: A typical amount of the medicine that you are given gram or milligram equivalents for. (For example, if you read an elixir bottle and it says 125 mg per 5 ml, the known dosage form is 5 ml.)

Dose ordered: This is the amount of grams or milligrams ordered.

Unknown amount to be given: This is what you are trying to determine—what amount of the medication should be given.

At times, the known unit on hand and the dose ordered may not be available in the same unit of measure (both in grams or both in milligrams). When this occurs, it is necessary to change the dose ordered to the same unit of measure as the known unit on hand (see chapter 8 for review of the metric system).

Example 1

The physician orders 0.5 g of guaifenesin (Robitussin) liquid q.4h for bronchial relief. The liquid is available in 125 mg per 5 ml. How many milliliters will you give the patient q.4h?

1. First, we must convert 0.5 g to mg (because the dose ordered must be in the same unit of measure as the known unit on hand).

    0.500 g = _____ mg

    kg hg dkg g dg cg mg   Move three places to the right.

    0.500 g = 500 mg

2. Next, complete the proportion.

    $$\frac{known\ unit\ on\ hand}{known\ dosage\ form} = \frac{dose\ ordered}{unknown\ amount\ to\ be\ given}$$

    $$\frac{125\ mg}{5\ ml} = \frac{500\ mg}{?\ ml}$$

    Cross multiply: $125 \times ? = 5 \times 500$
    $125 \times ? = 2500$

    Divide by 125: $\dfrac{125\ ?}{125} = \dfrac{2500}{125}$

    $? = 20$ ml

3. The solution is 20 ml every 4 hours (q.4h).

Example 2

The physician ordered Sinemet 300 mg t.i.d. Available is Sinemet tabs, 100 mg. How many tablets will you give?

1. There is no need for conversion—both what is needed and what is available are in mg.

2. $$\frac{known\ unit\ on\ hand}{known\ dosage\ form} = \frac{dose\ ordered}{unknown\ amount\ to\ be\ given}$$

    $$\frac{100\ mg}{1\ tab} = \frac{300\ mg}{?\ tabs}$$

    Cross multiply: $100 \times ? = 1 \times 300$
    $100\ ? = 300$

    Divide by 100: $\dfrac{100\ ?}{100} = \dfrac{300}{100}$

3. The solution is: $? = 3$ tabs, 3 times a day (t.i.d.).

Problems   Determine how much medicine to give in each of the following problems.

1) The physician ordered Diamox .25 g q.m. It is available in 125 mg per 5 ml. How many ml should be given?   _____

2) The physician ordered Sudafed 60 mg, t.i.d. It is available in 15 mg per 1 ml. How many ml should be given?   _____

3) The physician ordered Entamide 500 mg t.i.d. for 10 days. It is available in 1 g tablets. How many tablets should be given at each interval?   _____
What is the total number of tablets the patient will need over the 10-day period?   _____

4) The physician ordered Tylenol 650 mg q.4h. It is available in 50 mg per 1 ml syrup form. How many ml should be given?   _____

5) The physician ordered Bucladin-S Softabs 50 mg q.6h. They are available in 25 mg tablets. How many tablets should be given?   _____

6) The physician ordered 1500 mg Sofed with milk q.i.d. after meals and at bedtime. It is available in .5 g envelopes. How many envelopes should be used with milk at each interval?   _____
How many envelopes does the patient need a day?   _____

7) The physician ordered Valpin 50 mg t.i.d. It is available in .1 g caplets. How many caplets should be given?   _____

8) The physician ordered Inderal 30 mg q.i.d. It is available in 15 mg per 5 ml liquid. How many ml should be given?   _____

9) The physician ordered Motrin 250 mg q.6h. It is available in .5 g tablets. How many tablets should be given?   _____
How many will the patient need a day?   _____

10) The physician ordered Cortone 25 mg q.m. It is available in 12.5 mg per ml. How many ml should be given?   _____

Adult Parenteral Dosages

A **parenteral** medication is one that is injected into either the skin, a muscle, or a vein. These medications may be premeasured into single doses or may need to be measured from a multiuse container. The parenteral medications go to work more quickly than oral medications, so greater care must be taken to compute the dosages accurately.

We will again use the proportion method to determine how much medication to administer.

$$\frac{\text{known unit on hand}}{\text{known dosage form}} = \frac{\text{dose ordered}}{\text{unknown amount to be given}}$$

Example 1

The physician ordered Delalutin 150 mg IM. Available is Delalutin .45 g per ml. How many ml should be given?

1. First, we must convert 0.45 g to mg (because the dose ordered must be in the same unit of measure as the known unit on hand).

    0.450 g = _____ mg
    kg   hg   dkg   g   dg   cg   mg   Move three places to the right.
    0.450 g = 450 mg

2. Next, complete the proportion.

    $$\frac{\text{known unit on hand}}{\text{known dosage form}} = \frac{\text{dose ordered}}{\text{unknown amount to be given}}$$

    $$\frac{150\ mg}{1\ ml} = \frac{450\ mg}{?\ ml}$$

    Cross multiply: $150 \times ? = 1 \times 450$
    $150 \times ? = 450$

Divide by 150: $\dfrac{150\,?}{150} = \dfrac{450}{150}$

$? = 3\text{ ml}$

3. The solution is 3 ml intramuscularly.

Example 2

The physician ordered .5 g Robaxin IM. It is available in 50 mg per ml. How many ml should be given?

1. First, we must convert 0.5 g to mg (because the dose ordered must be in the same unit of measure as the known unit on hand).

   0.50 g = _____ mg

   kg   hg   dkg   g   dg   cg   mg   Move three places to the right.

   0.50 g = 500 mg

2. Next, complete the proportion.

   $$\dfrac{known\ unit\ on\ hand}{known\ dosage\ form} = \dfrac{dose\ ordered}{unknown\ amount\ to\ be\ given}$$

   $$\dfrac{50\ mg}{1\ ml} = \dfrac{500\ mg}{?\ ml}$$

   Cross multiply: $50 \times ? = 1 \times 500$

   $50 \times ? = 500$

   Divide by 150: $\dfrac{50\,?}{50} = \dfrac{500}{50}$

   $? = 10\text{ ml}$

Problems   Determine how much medicine to give in each of the following problems.

1) The physician ordered Solganal 10 mg IM the first week, 25 mg the second and third week, then 50 mg thereafter until a total of 1000 mg has been given. Solganal is available in .5 g per ml.

   a. How many ml should be given during week 1?   _____

   b. How many ml should be given during week 2?   _____

   c. How many ml should be given during week 3?   _____

   d. How many ml should be given during week 4?   _____

   e. What is the total number of weeks that the patient will need to receive a Solganal shot?   _____

2) The physician ordered Cortef 50 mg IM. It is available in 25 mg per ml. How many ml should be given?   _____

3) The physician ordered Streptomycin 75 mg IM. On hand is Streptomycin 100 mg per 5 ml. How much Streptomycin should be given?   _____

4) The physician ordered Permapen 1,000,000 units for deep IM injection. On hand is Permapen 750,000 units per 2 ml. What is the correct dosage to be administered?   _____

5) The physician ordered Dramamine 50 mg IM. It is available in 40 mg per 5 ml. What is the correct dosage? _____

6) The physician ordered .5 mg Lanoxin IV. It is available in 1 mg per 5 ml. How much Lanoxin should be given? _____

7) The physician ordered Epinephrine .5 mg IM. On hand is Epinephrine .01 mg per ml. How much Epinephrine should be given? _____

8) The physician ordered Ceporacin 1.5 g IM q.4h. On hand is Ceporacin 1 g per 5 ml. How many ml should be given? _____
How often? _____

9) The physician ordered Tobrex 3 mg per kg of patient weight IM. The patient weighs 75 kg. On hand is Tobrex 25 mg per ml. How many ml should be given? _____

10) The physician ordered Trobicin 2 g b.i.d. IM. On hand is Trobicin 500 mg per 2 ml. How many ml should be given? _____

Calculating IV Flow Rates

Intravenous (IV) fluids are widely used to replenish body fluids or nutrients lost, to give medication, or to keep a vein open for future use. Intravenous medications enter directly into the bloodstream and take immediate effect. This is the quickest, most efficient way to administer medication, but great care must be given to accuracy.

When an IV is used to replace fluids or give medication, the physician will order a **flow rate**. This is an indication of how fast the fluid should be entering the vein—it is essential to maintaining a certain prescribed amount over a period of time. The flow rate is given in milliliters per hour. It is your duty to determine the flow rate in drops (gtt) per minute. We will use the following formula to calculate IV flow rates:

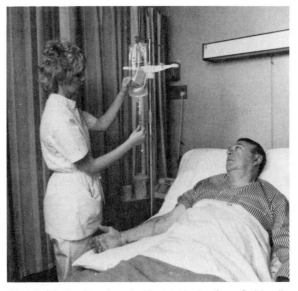

Figure 10–3 Reprinted with permission from Caldwell, *Nursing Assistant*, 5E, copyright 1989, Delmar Publishers Inc., Albany, NY.

$$\frac{(ml \div hr) \times calibrations\ (gtt\ /\ ml)}{minutes}$$

To use this formula, you need to know

the flow rate ordered by the physician

That is: the amount of solution ordered
the length of time that the solution is to be given

the calibrations (number of drops in a ml) for the specific solution you are using. Read the label with care. A **microdrop** delivers 60 gtt/ml—this is a standard measurement. A **macrodrop** is not standardized—it may be 10, 15, or 20 gtts/ml, depending on the manufacturer.

Example 1

The physician ordered 1000 ml D_5W (5% dextrose in water) to be infused over 10 hours. The infusion set is calibrated to deliver 10 gtt/ml.

$$\frac{(ml \div hr) \times calibrations}{minutes}$$

1. First, divide ml by hours (1000 ÷ 10 = 100).

2. Multiply that answer (100) by the calibrations (10 gtt/ml). 100 × 10 = 1000

3. Divide this answer (1000) by the number of minutes in 1 hour. 1000 ÷ 60 = 16.67 gtt/min

4. Round to the nearest whole drop. 16.67 rounds up to 17

Problems   Calculate the following IV flow rates.

1) The physician ordered D_5NS (5% dextrose in normal saline) 1000 ml to be infused over 8 hours. The infusion set is calibrated at 15 gtt/min.   _____

2) The physician ordered D_5W 2500 ml to be infused over 10 hours. The infusion set is calibrated at 20 gtt/min.   _____

3) The physician ordered D_5W 1000 ml to be infused over 24 hours. The infusion set is calibrated at 10 gtt/min.   _____

4) The physician ordered D_5NS 1000 ml to be infused over 10 hours. The infusion set is calibrated at 20 gtt/min.   _____

5) The physician ordered D_5W 1500 ml to be infused over 8 hours. The infusion set is calibrated at 15 gtt/min.   _____

Calculating Children's Dosages

Calculating children's dosages is different from figuring adult dosages because there are more factors involved. A child is defined as any human between infancy (not yet one year old) and puberty (the age when sexual reproduction is possible).

Children's dosages are calculated with regard to the child's height, weight, and age. There are five methods used to calculate children's medication dosages: Young's rule, Fried's rule, Clark's rule, body surface area (BSA), and per kilogram of body weight.

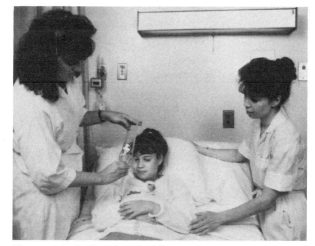

Figure 10–4   Reprinted with permission from Caldwell, *Nursing Assistant*, 5E, copyright 1989, Delmar Publishers, Albany, NY.

Below is a list of the methods used. Take note of which factors are involved in each rule (height, weight, and/or age).

Young's Rule

Young's rule uses the age of the child.

$$\frac{age\ of\ child}{age\ of\ child + 12} \times average\ adult\ dosage = child's\ dosage$$

Example 1

Decide how much Robitussin to give to a five-year-old patient if the adult dosage is normally 250 mg q.4h.

$$\frac{age\ of\ child}{age\ of\ child + 12} \times average\ adult\ dosage = child's\ dosage$$

$$\frac{5}{5 + 12} \times 250\ mg = 5 \times 250 \div 17 = 73.529$$

Round to 74 mg

Fried's Rule

Fried's rule is used to calculate dosages for children under two years of age. It may be used for older children as well.

$$\frac{age\ in\ months}{150} \times adult\ dose = infant\ dose$$

Example 1

Determine how much Amphojel to give a nine-month-old infant. The adult dosage is 10 ml q.6h.

$$\frac{age\ in\ months}{150} \times adult\ dose = infant\ dose$$

$$\frac{9}{150} \times 10\,ml = 9 \times 10 \div 150 = .6\,ml$$

Clark's Rule

Clark's rule uses the weight of the child (in pounds).

$$\frac{child's\ weight\ (lb)}{150} \times adult\ dose = child's\ dose$$

Example 1

What dosage will be given to an 80-lb boy when the adult dose is 50 mg?

$$\frac{child's\ weight\ (lb)}{150} \times adult\ dose = child's\ dose$$

$$\frac{80}{150} \times 50\,mg = 80 \times 50 \div 150 = 26.67\,mg$$

Round to 27 mg

Body Surface Area

The body surface area (BSA) is used to calculate dosages for children up to 12 years of age. It uses height and weight to determine the correct dosage amount. Because it takes both of these factors into consideration, it is considered to be one of the most accurate methods of calculating children's dosages (see figure 10–5).

To use the nomogram, draw a straight line from the child's weight to the child's height (which can be found in English or metric measurements). The point of intersection with the surface area chart tells you the child's body surface area. This information is used in the following formula:

$$\frac{BSA\ of\ child\ (m^2)}{1.7\,(m^2)} \times adult\ dose = child's\ dose$$

Example 1

Yvette Karr is a 17 lb, 30 in. infant (BSA = 0.4). The physician has ordered Tylenol for pain. The average adult dose of Tylenol is 500 mg. What dosage will be given to Yvette according to the BSA method?

$$\frac{BSA\ of\ child\ (m^2)}{1.7\,(m^2)} \times adult\ dose = child's\ dose$$

$$\frac{0.4}{1.7} \times 500\,mg = .4 \times 500 \div 1.7 = 117.65\,mg$$

Round to 118 mg.

Dosage per Kilogram of Body Weight

This rule uses the child's weight in kilograms.

1. Be sure that the weight is given in kilograms. (If it is given in pounds, divide the pounds by 2.2 to convert.)

2. Multiply the number of kilograms by the dosage prescribed for each kilogram.

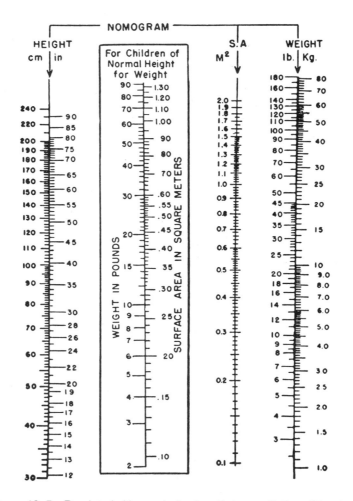

Figure 10–5 Reprinted with permission from Behrman, R.E. and Vaughan, V.C., *Nelson Textbook of Pediatrics*, 12th ed., 1983, W.B. Saunders Company, Philadelphia, PA 19105.

Example 1

Sudafed is administered to children with the dosage of 4 mg per kg daily in four divided doses. How much should an 88-lb child be given for an entire day? How much in each of the four doses?

1. 88 lb ÷ 2.2 = 40 kg

2. 4 mg × 40 kg = 160 kg a day

3. 160 divided into 4 equal doses = 40 mg per dose

Problems In the following problems, determine which rule or method is the most appropriate to use. Then, calculate the children's dosage.

1) Rosemary weighs 40 lb. She has been prescribed Motrin maximum strength. The maximum adult dose is 600 mg. How much should be given to Rosemary? _____

2) Jackie is a 30-lb, 42-in. girl (BSA = 0.62). She has been prescribed castor oil. The adult dose is 30 ml. How much should be given to Jackie? _____

3) The adult dosage of Demerol is 100 mg. How much should be given to an 11-year-old?   _____

4) The adult dosage of E-mycin is 250 mg q.6h. How much should be given to a 10-month-old child?   _____

5) The physician orders Sudafed be given to Abbey, who is a 55-lb girl. The adult dosage for Sudafed is 4 mg per kg body weight daily in four divided doses. How much should Abbey receive in one day?   _____
How much will each of the four doses contain?   _____

6) If the adult dose of Diamox is 1000 mg daily in 6 divided doses, what is the dosage for a 25-pound infant for the day?   _____
For each dose?   _____

7) The adult dosage for Dolobid is 1000 mg/day in 2 divided doses. What dosage should be given to a 60-lb child each day?   _____
In each of the two doses?   _____

Chapter 11

Vital Signs

Temperature Reading

Temperature can be read in **Fahrenheit** or **Celsius** degrees. Fahrenheit is the measurement most commonly used in the United States, but in most countries Celsius is preferred. Celsius measurement is used in the metric system.

The mercury thermometer was invented by Gabriel Fahrenheit. Fahrenheit established a scale for measuring temperature on a mercury column based on his finding that water froze at 32° and boiled at 212°. F is the abbreviation for Fahrenheit, and it is written after the temperature.

The Celsius scale was invented by Anders Celsius and is based on the system Fahrenheit had created, using the freezing and boiling points of water. Celsius decided to make the freezing point of water 0° and the boiling point 100°. Celsius measurement is sometimes called **centigrade**, meaning a system of 100 degrees or intervals. C is the abbreviation for Celsius.

Reading a patient's temperature can be done by several different methods. Rectal temperatures are the most accurate and take less time to provide an accurate reading. Oral temperatures are the next most accurate, but they take a bit longer to use. Axillary temperatures (temperature taken in the armpit or in the groin) are the least accurate and take much longer to yield a temperature reading.

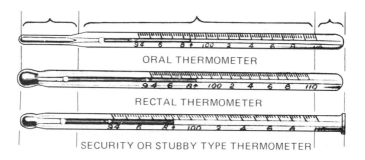

Figure 11–1a Reprinted with permission from Caldwell and Hegner, *Health Care Assistant*, copyright 1989, Delmar Publishers Inc., Albany, NY.

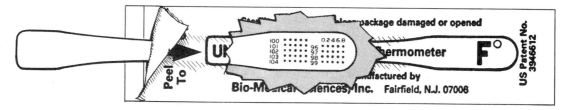

Figure 11–1b Reprinted with permission from Caldwell and Hegner, *Health Care Assistant*, copyright 1989, Delmar Publishers Inc., Albany, NY.

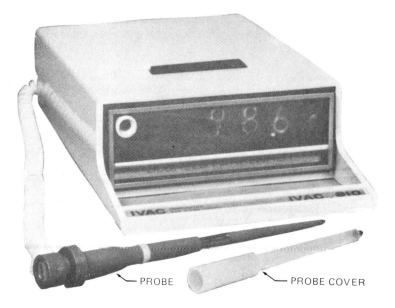

PROBE   PROBE COVER

Figure 11–1c Reprinted with permission from Keir, *Medical Assisting*, 2E, copyright 1989, Delmar Publishers Inc., Albany, NY.

The normal range for body temperature is 97° to 100° Fahrenheit (F) and 36.1° to 37.8° Celsius (C).

Rectal: average body temperature is 99.6°F or 37.6°C; leave thermometer in place 2 minutes.

Oral: average body temperature is 98.6°F or 37°C; leave thermometer in place 3 minutes.

Axillary: average body temperature is 97.6°F or 36.4°C; leave thermometer in place 10 minutes.

Rectal temperature is the most accurate because it is taken internally and most closely represents the patient's internal body temperature. That is why it is the highest of the three temperatures. Axillary temperatures, on the other hand, are taken in a fold of the body (externally) and yield the lowest body temperature.

Temperatures can be taken with a conventional glass thermometer or with an electronic thermometer. The glass thermometer uses mercury, which expands with heat. As the patient's temperature rises, the mercury expands and fills the glass column. The temperature is read at the highest point of mercury. The scale up the side divides each degree into five parts, each representing two-tenths. The long lines represent whole degrees, and the shorter lines represent the two-tenth intervals (see figure 11–2).

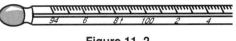

Figure 11–2

To read a thermometer, it is important to find out what whole degree has been reached and then read the two-tenth intervals that follow. For example, the mercury column has reached 98 degrees and two intervals. Since each interval represents two-tenths, this thermometer reads 98.4 (see figure 11–3).

Figure 11–3   98⁴

In the next example, the mercury column has reached a long line that is not labeled. It is between 98 and 100, so it must represent 99 degrees. (The odd degree marks are often not labeled.) It is three intervals past 99, so this thermometer reads 99.6 (see figure 11–4).

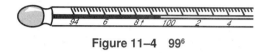

Figure 11–4   99⁶

The temperature should always be read in these even two-tenth intervals (e.g., 98.**6**, 99.**8**, 100.**2**, 96.**0**, etc.) and not in odd tenths of a degree (e.g., 98.**7** or 100.**3**). The temperature is usually written with the degree as a whole number and the two-tenths interval as a superscript above and to the right of the degree (example: 98.6 should be written 98⁶). This eliminates errors that sometimes occur when we read decimals.

The electronic thermometer gives a digital reading of the temperature using the two-tenths intervals. It is easier to read and quicker to use than the conventional thermometer. It is also easier to use from a hygienic standpoint—disposable protective covers are used to avoid the transfer of germs between patients.

Problems   Complete the following sentences with the correct information.

1) There are two types of thermometers most commonly used: _____ and _____.

2) The normal body temperature varies depending on the method used. The normal body temperature taken rectally is _____F or _____C. The normal body temperature taken orally is _____F or _____C. The normal body temperature taken axillary is _____F or _____C. The _____ method is most accurate and the _____ method is least accurate.

3) The body temperature 98.0 should be written _____.

4) The rectal method of taking body temperature is the most accurate because it is taken _____.

5) The _____ thermometer is easier to use from a hygienic standpoint because disposable protective covers are used.

In problems 6–20, read the following thermometers and write the temperature correctly (figure 11–5).

6) _____

7) _____

8) _____

9) _____

10) _____

Figure 11–5

(figure 11–6)

11) _____

12) _____

13) _____

14) _____

15) _____

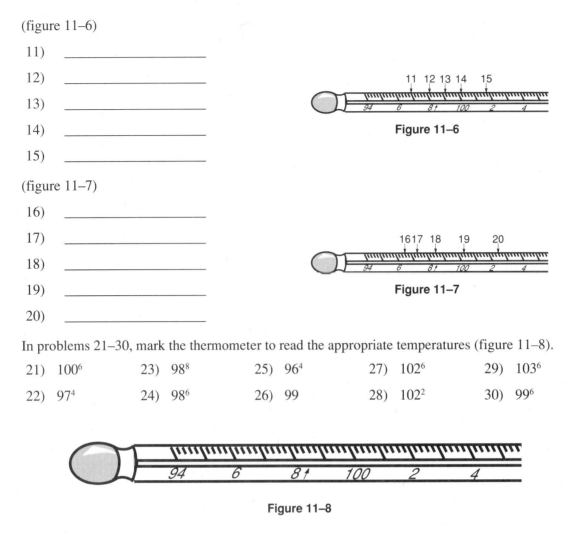

Figure 11–6

(figure 11–7)

16) _____

17) _____

18) _____

19) _____

20) _____

Figure 11–7

In problems 21–30, mark the thermometer to read the appropriate temperatures (figure 11–8).

21) 100^6 23) 98^8 25) 96^4 27) 102^6 29) 103^6

22) 97^4 24) 98^6 26) 99 28) 102^2 30) 99^6

Figure 11–8

Temperature Conversion

Because there are two systems that are used to measure temperature, it occasionally becomes necessary to convert from one system to the other. To convert from Celsius to Fahrenheit or Fahrenheit to Celsius, we will set up a proportion. But before we can set up a proportion, we must better understand the two systems (see figure 11–9).

There is 180° between the freezing point and the boiling point in the Fahrenheit system. There is only 100° between the freezing and boiling points in the Celsius system. This is a ratio of 180:100. This can be reduced to 9:5.

It is also important to notice that there is 32° difference between the systems at the beginning of the mercury columns (freezing point). If we could take 32° away from Fahrenheit, they would be similar systems. So, F – 32 = C or (F – 32):C.

Thus, we have two ratios that we can use to write a proportion.

$$\frac{9}{5} = \frac{(F - 32)}{C}$$ Fahrenheit information on top.
                                          Celsius information on bottom.

Any temperature that we wish to convert can be put into this proportion. This will allow conversion from one system to the other.

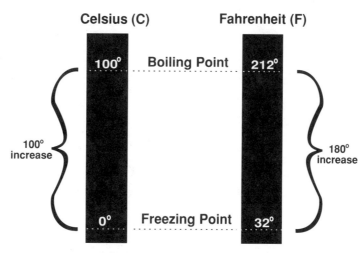

Figure 11–9
1. Celsius freezes at 32° lower than Fahrenheit.
2. Then Celsius temperature increases 100° to boiling compared to Fahrenheit's 180° increase (100 to 180 → $\frac{100}{180} = \frac{5}{9}$ reduced).
3. $C = (F - 32) \times \frac{5}{9}$ or $F = \frac{9 \times C}{5} + 32$

Example 1

Convert 100°F to Celsius.

$\frac{9}{5} = \frac{(F - 32)}{C}$ Begin with the proportion of Fahrenheit to Celsius.

$\frac{9}{5} = \frac{(100 - 32)}{C}$ Put 100° in for the F symbol.

$\frac{9}{5} = \frac{(68)}{C}$ Complete all work in parentheses before cross multiplying.

$9C = 5 \times 68$ Cross multiply.

$\frac{9C}{9} = \frac{340}{9}$ Divide both sides by 9 to get the C alone.

$C = 37.\overline{7}$, rounds to 37.8°C

Example 2

Convert 46°C to Fahrenheit.

$\frac{9}{5} = \frac{(F - 32)}{C}$ Begin with the proportion of Fahrenheit to Celsius.

$\frac{9}{5} = \frac{(F - 32)}{46}$ Put 46° in for the C symbol.

$9 \times 46 = 5 (F - 32)$ Cross multiply.

$414 = 5F - 160$ Multiply everything in the parentheses by 5.

$414 + 160 = 5F - 160 + 160$ Add 160 to both sides to get the 5F alone.

$\frac{574}{5} = \frac{5F}{5}$ Divide both sides by 5 to get the F alone.

$114.8 = F$

Problems   Convert the following temperatures from the system they are in to the other system.

1) 56.7°C   _____   11) 98.6°F   _____

2) 11°C   _____   12) 102.2°F   _____

3) 46.4°F   _____   13) 100°C   _____

4) 86°F   _____   14) 20°C   _____

5) 44°C   _____   15) 109°F   _____

6) 33.8°F   _____   16) 93°F   _____

7) 15°C   _____   17) 38°C   _____

8) 131°F   _____   18) 0°C   _____

9) 106°F   _____   19) 12°C   _____

10) 6°F   _____   20) 84°C   _____

21) Janice took a patient's temperature with a Celsius thermometer. The temperature read 37.5°C. Is this in the normal body temperature range (show work)?   _____

22) Rachel took a patient's temperature with a Fahrenheit thermometer. The temperature read 103°F. What is the temperature in Celsius?   _____

Recording Temperature/Pulse/Respiration

A pulse is the number of contractions the heart makes in one minute. It is found by placing pressure over the radial artery in the wrist of the patient and counting the number of pulsations that you feel in the blood vessel. To find the pulse for one minute, it is most common to either find the pulse for 30 seconds ($\frac{1}{2}$ a minute) and then double that number or to find the pulse for 10 seconds ($\frac{1}{6}$ a minute) and then multiply that number by 6. A patient's pulse is usually taken while you are taking his temperature to save time and to eliminate the distraction of a conversation while you are counting.

A patient's respiration is a full cycle of inhalation and exhalation of breath. It is often necessary to count the number of respirations in one minute. This should be done without the patient's knowledge, because breathing can be controlled.

Temperature, pulse, and respiration are often recorded together on one graphic sheet. The graphic sheet we will use is similar to a sheet of graph paper with horizontal and vertical lines drawn through it. The date and times are written at the top, and a temperature scale is written along the side, ranging from a low temperature of 96° to a high of 105°. Beneath this, there are boxes for recording the patient's pulse and respiration. The appearance of the graphic sheet may change from place to place, but the basic information will be the same (see figure 11–10).

The temperature is recorded by placing a dot in the appropriate time space. The dot should be centered between the vertical lines, not on a line, so it is clear which time it represents. The appro-

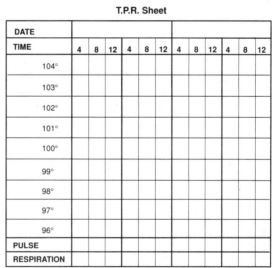

Figure 11–10

T.P.R. Sheet

DATE												
TIME	4	8	12	4	8	12	4	8	12	4	8	12
104°												
103°												
102°												
101°												
100°												
99°												
98°												
97°												
96°												
PULSE												
RESPIRATION												

priate temperature is indicated by placing the dot either on the appropriate temperature line or placing it proportionately between two lines. For example, 98⁶ would be placed between 98 and 99—closer to 99. 100² would be placed between 100 and 101—closer to 100. For example, the dot below represents 103⁸ (see figure 11–11):

T.P.R. Sheet

DATE												
TIME	4	8	12	4	8	12	4	8	12	4	8	12
104°												
103°	*											

Figure 11–11

Because normal body temperature is different depending on the method used to take the temperature (oral, rectal, or axillary), it is customary to use symbols on the chart to denote which method was used. Axillary temperatures should have the abbreviation A or Ax written beside the dot and circled. Rectal temperatures should have an R written beside the dot and circled. Oral temperatures do not need a symbol. It is assumed that if no symbol is used, the temperature was taken orally.

When you begin a TPR sheet for a patient and the first dot is drawn, it is helpful to draw a short line from the dot back to the left. This makes the beginning of the graphic line and draws your attention to the first dot. Then when a second dot is drawn, it should be connected back to the previous dot to make a continuous graph line. Always use the same color pen that you used to make the last dot to draw the connecting line back to the previous dot. If TPR is not taken every four hours, be careful to skip over the unused time slots.

The pulse and respiration are not recorded graphically like the temperature. Instead, they are written numerically in the boxes at the bottom of the graphic sheet.

The appropriate colored pens should be used to record TPR:

blue or black	7 am to 3 pm
green	3 pm to 11 pm
red	11 pm to 7 am

Example 1

Chart the following information on a blank TPR graphic sheet (see figure 11–12).

Date	Time	Temperature	Pulse	Respiration
3/5/90	4:00 am	103⁶ R	126	38
	8:00 am	103² R	120	36
	12:00 noon	101⁸ R	116	30
	4:00 pm	100⁴ R	110	30
	8:00 pm	101 R	114	32

Problems Chart the following information on a blank TPR graphic sheet.

1) Patient: Allison Hamilton   Room: 362–67
   Physician: Dr. E. J. Peters

Date	Time	Temperature	Pulse	Respiration
3/5/90	4:00 am	103⁶ R	126	38
	8:00 am	103² R	120	36
	12:00 noon	101⁸ R	116	30
	4:00 pm	100⁴ R	110	30

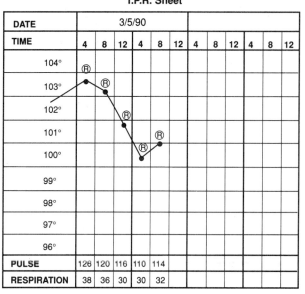

Figure 11–12

Date	Time		Temp	Pulse	Resp
	8:00	pm	101 R	114	32
	12:00	M.N.	102^8 R	116	30
3/6/90	4:00	am	101^6 R	104	26
	8:00	am	In OR—emergency surgery		
	12:00	noon	100^4 Ax	120	28
	4:00	pm	99^2 Ax	108	24
	8:00	pm	98^6	96	20
	12:00	M.N.	Sleeping—not disturbed		
3/7/90	4:00	am	98^6	78	16
	8:00	am	99^4	80	18
	12:00	noon	100	84	16
	4:00	pm	99^4	82	16
	8:00	pm	101^2	90	22
	12:00	M.N.	102^8 R	108	30
3/8/90	4:00	am	103^6 R	116	34
	8:00	am	103 R	120	32
	12:00	noon	99^4	108	24
	4:00	pm	99^6	104	20
	8:00	pm	99^8	102	18
3/9/90	8:00	am	100	96	18
	12:00	noon	99^2	84	14
	4:00	pm	98^4	78	10
	8:00	pm	98^6	72	12
3/10/90	8:00	am	98^6	64	10
	12:00	noon	99^2	76	12
	8:00	pm	98^6	82	10
3/11/90	8:00	am	99	70	14
	4:00	pm	98^8	68	12
3/12/90	8:00	am	Discharged		

T.P.R. Sheet

DATE												
TIME	4	8	12	4	8	12	4	8	12	4	8	12
104°												
103°												
102°												
101°												
100°												
99°												
98°												
97°												
96°												
PULSE												
RESPIRATION												
DATE												
TIME	4	8	12	4	8	12	4	8	12	4	8	12
104°												
103°												
102°												
101°												
100°												
99°												
98°												
97°												
96°												
PULSE												
RESPIRATION												

T.P.R. Sheet

DATE												
TIME	4	8	12	4	8	12	4	8	12	4	8	12
104°												
103°												
102°												
101°												
100°												
99°												
98°												
97°												
96°												
PULSE												
RESPIRATION												
DATE												
TIME	4	8	12	4	8	12	4	8	12	4	8	12
104°												
103°												
102°												
101°												
100°												
99°												
98°												
97°												
96°												
PULSE												
RESPIRATION												

2) Patient: Jack Warner   Room: 822–99
Physician: Dr. Wm. Honigsford

Date	Time		Temperature	Pulse	Respiration
6/2/91	4:00	am	102^6 R	118	38
	8:00	am	103^2 R	116	40
	12:00	noon	102^8 R	118	36
	4:00	pm	100^4	120	30
	8:00	pm	101	116	36
	12:00	M.N.	102^4	110	30
6/3/91	4:00	am	103^2	104	38
	8:00	am	102^8 R	110	32
	12:00	noon	100^4 R	116	28
	4:00	pm	99^2	104	24
	8:00	pm	101^4	108	30
	12:00	M.N.	100^6	112	28

6/4/91	4:00	am	100	100	32
	8:00	am	99^4	88	28
	12:00	noon	100	90	24
	4:00	pm	100^8	88	20
	8:00	pm	101^2	90	22
	12:00	M.N.	103^8	118	36
6/5/91	4:00	am	102^6 R	116	34
	8:00	am	103 R	118	28
	12:00	noon	100^4 R	108	22
	4:00	pm	99^6 R	102	20
	8:00	pm	98^8	102	18
6/6/91	8:00	am	100	96	18
	12:00	noon	99^2	84	14
	4:00	pm	98^6	80	16

T.P.R. Sheet

DATE												
TIME	4	8	12	4	8	12	4	8	12	4	8	12
104°												
103°												
102°												
101°												
100°												
99°												
98°												
97°												
96°												
PULSE												
RESPIRATION												
DATE												
TIME	4	8	12	4	8	12	4	8	12	4	8	12
104°												
103°												
102°												
101°												
100°												
99°												
98°												
97°												
96°												
PULSE												
RESPIRATION												

T.P.R. Sheet

DATE												
TIME	4	8	12	4	8	12	4	8	12	4	8	12
104°												
103°												
102°												
101°												
100°												
99°												
98°												
97°												
96°												
PULSE												
RESPIRATION												
DATE												
TIME	4	8	12	4	8	12	4	8	12	4	8	12
104°												
103°												
102°												
101°												
100°												
99°												
98°												
97°												
96°												
PULSE												
RESPIRATION												

3) Patient: Phyllis Pernell   Room: 6245
   Physician: Dr. Phillip Joshid

Date	Time		Temperature	Pulse	Respiration
11/11/91	8:00	pm	99^2	96	22
	12:00	M.N.	102^6	88	24
11/12/91	8:00	am	101^4 R	90	20
	12:00	noon	102^6 R	106	26
	4:00	pm	103^6 R	120	34
	8:00	pm	102^8 R	124	38
	12:00	M.N.	103^6 R	136	34
11/13/91	4:00	am	104^2 R	128	34
	8:00	am	101^8 R	130	30
	12:00	noon	100^6 R	122	24
	4:00	pm	99 R	96	28
	8:00	pm	100^2	100	26
11/14/91	4:00	am	100	88	22
	8:00	am	99^2	76	16

T.P.R. Sheet

DATE												
TIME	4	8	12	4	8	12	4	8	12	4	8	12
104°												
103°												
102°												
101°												
100°												
99°												
98°												
97°												
96°												
PULSE												
RESPIRATION												
DATE												
TIME	4	8	12	4	8	12	4	8	12	4	8	12
104°												
103°												
102°												
101°												
100°												
99°												
98°												
97°												
96°												
PULSE												
RESPIRATION												

T.P.R. Sheet

DATE												
TIME	4	8	12	4	8	12	4	8	12	4	8	12
104°												
103°												
102°												
101°												
100°												
99°												
98°												
97°												
96°												
PULSE												
RESPIRATION												
DATE												
TIME	4	8	12	4	8	12	4	8	12	4	8	12
104°												
103°												
102°												
101°												
100°												
99°												
98°												
97°												
96°												
PULSE												
RESPIRATION												

	Time	Temperature	Pulse	Respiration
	12:00 noon	99^8	68	18
	4:00 pm	100^8	72	22
	8:00 pm	102^6	90	26
	12:00 M.N.	102^8 R	94	22
11/15/91	4:00 am	103 R	100	26
	8:00 am	104^2 R	112	38
	12:00 noon	102^6 R	118	40
	4:00 pm	99^8 R	102	32
	8:00 pm	Emergency appendectomy		
11/16/91	4:00 am	99^6 Ax	88	30
	8:00 am	98^6	74	24
	12:00 noon	98^2	68	18
	4:00 pm	98^6	50	16
11/17/91	8:00 am	99	42	14
	4:00 pm	98^6	36	12
11/18/91	8:00 am	98^4	32	16

4) Patient: Donald Richmond   Room: 227
Physician: Dr. Asmin Rashan

Date	Time	Temperature	Pulse	Respiration
9/19/90	8:00 am	104^6 R	126	30
	12:00 noon	104^6 R	120	26
	4:00 pm	103^2 R	116	22
	8:00 pm	103^8 R	120	28
	12:00 M.N.	102^8 R	112	20
9/20/90	4:00 am	99^8 R	106	20
	8:00 am	In OR—partial colostomy		
	12:00 noon	100^2 Ax	96	12
	4:00 pm	99^6 Ax	92	10
	8:00 pm	99^8 Ax	84	14
	12:00 M.N.	100	76	10
9/21/90	4:00 am	102^6	88	14
	8:00 am	101^2	78	10
	12:00 noon	100^8	70	16
	4:00 pm	99	62	18
	8:00 pm	100^2	70	18
	12:00 M.N.	100^8	66	16
9/22/90	4:00 am	101	60	18
	8:00 am	101^8	72	20
	12:00 noon	101^8	76	24
	4:00 pm	102	84	26
	8:00 pm	99^6	88	24
9/23/90	8:00 am	99^2	78	18
	4:00 pm	98^6	66	16
	8:00 pm	98^6	54	14
9/24/90	8:00 am	98^6	48	12

T.P.R. Sheet

DATE												
TIME	4	8	12	4	8	12	4	8	12	4	8	12
104°												
103°												
102°												
101°												
100°												
99°												
98°												
97°												
96°												
PULSE												
RESPIRATION												
DATE												
TIME	4	8	12	4	8	12	4	8	12	4	8	12
104°												
103°												
102°												
101°												
100°												
99°												
98°												
97°												
96°												
PULSE												
RESPIRATION												

T.P.R. Sheet

DATE												
TIME	4	8	12	4	8	12	4	8	12	4	8	12
104°												
103°												
102°												
101°												
100°												
99°												
98°												
97°												
96°												
PULSE												
RESPIRATION												
DATE												
TIME	4	8	12	4	8	12	4	8	12	4	8	12
104°												
103°												
102°												
101°												
100°												
99°												
98°												
97°												
96°												
PULSE												
RESPIRATION												

5) Patient: Gretta Rauel   Room: 8245–0
   Physician: Dr. Joseph Wallace

Date	Time	Temperature	Pulse	Respiration
12/23/90	8:00 pm	104^2	126	36
	12:00 M.N.	104^6	116	34
12/24/90	8:00 am	104^6 R	120	36
	12:00 noon	103^6 R	116	34
	4:00 pm	103^2 R	120	34
	8:00 pm	102^6 R	114	38
	12:00 M.N.	102^4 R	116	36
12/25/90	4:00 am	101^6 R	118	34
	8:00 am	101^2 R	120	30
	12:00 noon	100^8 R	112	34
	4:00 pm	100^6 R	100	30
	8:00 pm	99^2	88	26

12/26/90	4:00 am	100	88	22
	8:00 am	99^8	64	16
	12:00 noon	99^8	68	18
	4:00 pm	98^2	72	14
	8:00 pm	100^8	66	20
	12:00 M.N.	98^8	70	22
12/27/90	4:00 am	98^6	72	18
	8:00 am	101^2	88	20
	12:00 noon	102^6 R	92	24
	4:00 pm	99^8 R	78	18
	8:00 pm	98	70	22
12/28/90	4:00 am	99^6	66	18
	8:00 am	98^6	74	18
	12:00 noon	98^2	68	18
	4:00 pm	98^6	50	16
12/29/90	8:00 am	99	42	14
	4:00 pm	98^6	36	12

T.P.R. Sheet

DATE												
TIME	4	8	12	4	8	12	4	8	12	4	8	12
104°												
103°												
102°												
101°												
100°												
99°												
98°												
97°												
96°												
PULSE												
RESPIRATION												
DATE												
TIME	4	8	12	4	8	12	4	8	12	4	8	12
104°												
103°												
102°												
101°												
100°												
99°												
98°												
97°												
96°												
PULSE												
RESPIRATION												

T.P.R. Sheet

DATE												
TIME	4	8	12	4	8	12	4	8	12	4	8	12
104°												
103°												
102°												
101°												
100°												
99°												
98°												
97°												
96°												
PULSE												
RESPIRATION												
DATE												
TIME	4	8	12	4	8	12	4	8	12	4	8	12
104°												
103°												
102°												
101°												
100°												
99°												
98°												
97°												
96°												
PULSE												
RESPIRATION												

Reading a Sphygmomanometer

As blood is pumped from the heart, it creates pressure on the blood vessels. The pressure when the heart is contracting is called the **systolic** pressure. The pressure between contractions decreases, and it is called the **diastolic** pressure.

To find a patient's blood pressure, it is necessary to know how to read a sphygmomanometer. There are two types of sphygmomanometers—one with a mercury column display similar to a thermometer and one with a dial display, called an **aneroid** sphygmomanometer.

The mercury column is divided into increments of 10, represented by the longer lines. Within each increment, there are smaller increments of two, represented by the shorter lines. To read the column, find where the mercury level is, determine which long line it is past and then count by twos until you reach the short line.

For example, on the mercury column below, the mercury line is past the long line representing 120. Counting by twos, we find that it reads 126 (see figure 11–13).

The aneroid display looks like the dial on bathroom scales. There is a long needle in the center that moves clockwise as the pressure in the blood pressure cuff increases. Like the mercury column, it is also divided into increments of 10, represented by longer lines. Within each increment, there are smaller increments of two, represented by shorter lines. Every other long line is labeled beginning with 20, 40, and so forth. The odd increments of ten (30, 50, etc.) are not labeled. To read the dial, determine where the needle is pointing. Next, decide which long line it is past, then count by twos until you reach the short line (see figure 11–14).

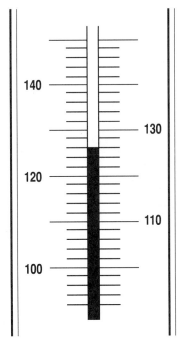

Figure 11–13

Figure 11–14

Problems   Read the following sphygmomanometers and record your answers in the spaces provided.

(figure 11–15)

1) _____

2) _____

3) _____

4) _____

5) _____

6) _____

7) _____

8) _____

9) _____

10) _____

11) _____

12) _____

(figure 11–16)

13) _____

14) _____

15) _____

16) _____

17) _____

18) _____

19) _____

20) _____

21) _____

22) _____

23) _____

24) _____

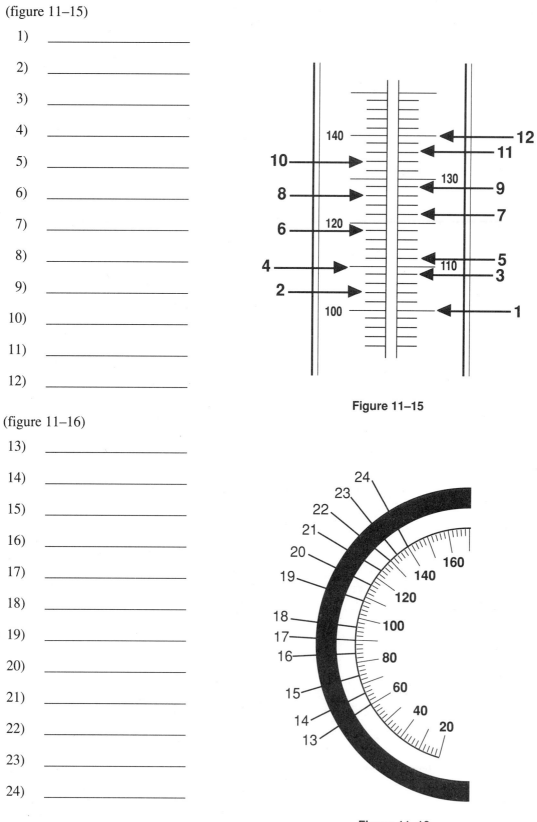

Figure 11–15

Figure 11–16

Recording Blood Pressure

To record blood pressure, both the systolic and diastolic pressures should be recorded. They are recorded with two continuous graphed lines that look similar to the temperature graph. The systolic pressure is always higher than diastolic, so the two lines can share the same graph. The blood pressure chart looks like a piece of graph paper with the date and times labeled across the top. Dots are drawn in a similar way to the temperature chart, one for systolic pressure and one for diastolic. Then, when the next blood pressure reading is recorded, connecting lines are drawn back to the previous dots—systolic connecting in one line above and diastolic connecting in one line below. Appropriate colored pens should be used for each shift (see figure 11–17).

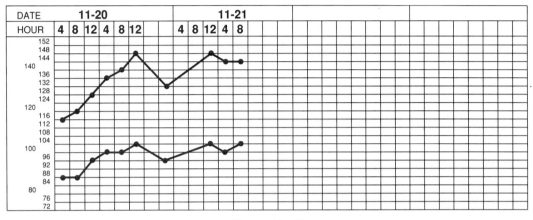

Figure 11–17

Problems Record the following patients' blood pressures.

1) Stacy McFarland was admitted to Boston Medical Center on November 12. Record her blood pressures as given below (figure 11–18).

November 12	Noon	188/98			M.N.	140/90
	4 pm	196/100		November 14	8 am	126/84
	8 pm	178/86			noon	120/92
	M.N.	174/82			4 pm	116/84
November 13	4 am	140/80			8 pm	110/80
	8 am	126/74		November 15	8 am	120/80
	noon	120/82			4 pm	128/82
	4 pm	166/98		November 16	8 am	120/80
	8 pm	162/94				

2) Record the following blood pressure reading for Sarah Byers during her stay at Hudson Memorial Hospital (figure 11–18).

June 11	4 am	106/84			8 pm	140/100
	8 am	98/68			M.N.	128/90
	noon	98/74		June 13	4 am	148/110
	4 pm	110/80			8 am	166/124
	8 pm	104/68			noon	180/128
June 12	8 am	120/76			4 pm	170/120
	noon	134/84			8 pm	156/114
	4 pm	146/92			M.N.	150/108

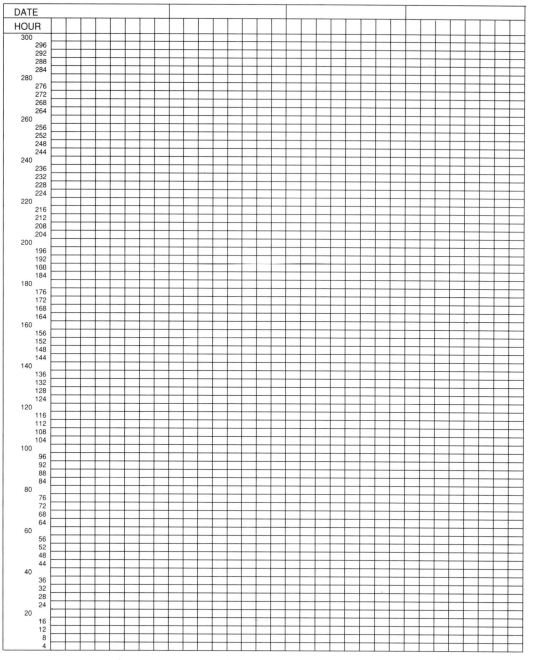

Figure 11–18

June 14	4 am	144/96	June 15	8 am	120/84
	8 am	136/94		noon	130/90
	noon	in x-ray		4 pm	140/94
	4 pm	120/80		8 pm	136/86
	8 pm	136/86		M.N.	128/82
	M.N.	120/84	June 16	8 am	120/80
				noon	discharged

Chapter 12

Intake and Output

When patients are hospitalized or under supervised care, it may at times be necessary to place them on **intake and output**. This is a process where the patients' fluid intake into their bodies and output from their bodies are recorded. Patients placed on intake and output often have digestive disorders or are in critical condition. The process is ordered by a physician and is not common practice unless authorized.

To record intake and output, it is first important to understand the various ways that fluids can be taken in or expelled from the body. Intake occurs in the following ways:

- orally
- intravenously
- by irrigation

Output occurs in the following ways:

- bowel movements (BM)
- emesis (vomiting)
- urine
- irrigation drainage

Solid bowel movements are not considered when totaling the amount of output, but they are usually noted on the chart with a brief description.

When recording intake and output, the fluid must be measured or approximated. The amount is recorded in cubic centimeters (cc), which are the same as milliliters (see chapter 8, on the metric system).

Figure 12–1 Reprinted with permission from Caldwell, *Nursing Assistant*, 5E, copyright 1989, Delmar Publishers Inc., Albany, NY.

Intake and output are recorded each hour of the day and are totaled at the end of each work shift and again at the end of the day (a day is considered to be 7 am to 7 am). As with other medical charts, appropriately colored pens should be used for different shifts.

7 am to 3 pm	blue or black ink
3 pm to 11 pm	green ink
11 pm to 7 am	red ink

Below is a table of the various types of intake and output and instructions on how to measure each type accurately.

- Oral intake is measured by determining what size dish the item was served from. At the top of most intake and output charts is a table of measurement for various dishes, cups, and bowls (example: 1 water glass = 180 cc). Occasionally, it will also be necessary to refer to the table of metric-English equivalents found in chapter 8.

  The common measurements are as follows:

large glass	240 cc		coffeepot	240 cc
water glass	180 cc		small bowl	120 cc
juice glass	100 cc		soup bowl	200 cc
cup	120 cc			

- IV intake is measured by determining how much has dripped from the bag over a given period of time. (Example: If you started with 1000 cc and 850 cc are left, the patient's intake is 150 cc—arrived at by subtraction.)
- Irrigation is determined by actual measurement of the liquid entering the irrigation tube.
- Liquid bowel movements are measured by approximating the amount of the liquid. Often, patients on intake and output will use a bedpan, and the amount will be easily determinable.
- Emesis amount is determined by actual measurement of the liquid.
- Urine can be determined in one of two ways. If the patient has a catheter, the actual amount of urine can be measured when the bag is emptied. If the patient is not on a catheter, the amount should be measured after each urination.
- Irrigation drainage is determined by actual measurement of the liquid being expelled from the irrigation tube.

On the following pages, you will find exercises involving patients who have been put on intake and output. It is your responsibility to record their intake and output on the charts given.

The chart is divided into two parts: intake on the left side and output on the right side. There is space provided on each side for remarks.

On the chart, the time of day is printed along the left side in hourly increments. Any intake or output during each hour is added together and written in the appropriate box. For example, if a patient eats at 7:30 am, that intake is recorded in the 7:00 am block along with any other intake during the 7:00 hour.

At the end of each eight-hour shift, it is the nurse on duty's responsibility to total all of the columns for her shift. The night nurse (11 pm–7 am shift) also has the responsibility of totaling the entire day's figures at the bottom and starting a new day's sheet.

Problems   Complete each intake and output sheet on the forms provided.

1) Jeff Kerchoff was hospitalized January 6 for high fever and vomiting. An IV was started. Initially, he was allowed clear liquids. As he stabilized, a soft diet was introduced. Record the following on an I/O sheet. Total all eight-hour and end-of-day columns. Use the correct color for each shift.

January 6   7:45   Brought into emergency room

             8:00   Vomited 400 cc clear liquid

             8:10   IV of 10% glucose started

             8:45   Water sips—2 Tbl

             9:00   Admitted to hospital for tests

             9:05   Ice chips—5 cubes (1 cube = 5 cc)

             9:10   Liquid bowel movement—300 cc greenish liquid

             9:15   Vomited 50 cc yellowish liquid

           10:00   A second IV of Emetrol started

           10:25   Urine output of 250 cc

           10:30   Water sips—4 Tbl

           11:00   Liquid bowel movement—200 cc

           11:45   Water sips—$\frac{1}{2}$ juice glass

           12:00   IV 1 had absorbed 350 cc since 8:00

           12:25   Vomited 100 cc clear liquid

           12:30   Emetrol IV taken out. Had absorbed 150 cc since 10:00

             1:00   Urine output of 100 cc

             1:15   Ice chips—5 cubes

             1:30   Drinks 2 large glasses of water as part of a medical test

             1:45   Vomited 300 cc clear liquid

             3:00   IV 1 had absorbed 400 cc since 12:00

             4:30   Water sips—4 Tbl

             5:00   Urine output of 250 cc

             7:30   1 formed greenish stool

             8:15 $\frac{1}{2}$ juice glass of ginger ale

           10:00   IV 1 had absorbed 350 cc since 3:00

           10:45   1 glass of water

January 7 12:15   Urine output of 300 cc

             4:00   Ice chips—10 cubes

             6:30   IV 1 had absorbed 400 cc since 10:00

             7:45 $\frac{1}{2}$ juice glass of ginger ale

             8:00   Urine output of 200 cc

           10:30   Small bowl chicken broth

           12:00   Large glass of water

                      Small bowl of liquid Jello

                      $\frac{1}{2}$ small bowl of chicken broth

             2:00   1 formed brown stool

2) Alvin Topplin has been admitted to Covington General Hospital as a result of a car accident. Due to internal injuries, he has been placed on intake and output. Part of his bowel has had to be reconstructed. A Levine tube is in place, and he is on a 10% glucose IV and also is beginning a soft diet.

August 14 7:00   Breakfast: $\frac{1}{2}$ small bowl Jello

                                1 glass of apple juice

                                1 glass of water

             7:45   IV had absorbed 300 cc

             7:50   Urine output of 450 cc

             8:00   Levine tube irrigated with normal saline 15 cc

             8:10   Ice chips—5 cubes

Large glass—240 cc   Coffee pot—240 cc
Water glass—180 cc   Small bowl—120 cc
Juice glass—100 cc   Soup bowl—200 cc
Cup—120 cc

Time	Oral	I.V.	Irrig	Remarks	B.M.	Emesis	Urine	Irrig	Remarks
7:00									
8:00									
9:00									
10:00									
11:00									
12:00									
1:00									
2:00									
Shift Total									
3:00									
4:00									
5:00									
6:00									
7:00									
8:00									
9:00									
10:00									
Shift Total									
11:00									
12:00									
1:00									
2:00									
3:00									
4:00									
5:00									
6:00									
Shift Total									
Grand Total									

Large glass—240 cc   Coffee pot—240 cc
Water glass—180 cc   Small bowl—120 cc
Juice glass—100 cc   Soup bowl—200 cc
Cup—120 cc

Time	Oral	I.V.	Irrig	Remarks	B.M.	Emesis	Urine	Irrig	Remarks
7:00									
8:00									
9:00									
10:00									
11:00									
12:00									
1:00									
2:00									
Shift Total									
3:00									
4:00									
5:00									
6:00									
7:00									
8:00									
9:00									
10:00									
Shift Total									
11:00									
12:00									
1:00									
2:00									
3:00									
4:00									
5:00									
6:00									
Shift Total									
Grand Total									

	8:30	Liquid bowel movement—250 cc greenish liquid
	10:15	Irrigation drainage 200 cc—light brown liquid
	11:20	$\frac{1}{2}$ large glass of ginger ale
	11:45	Urine output of 200 cc
	12:00	Levine tube irrigated with normal saline 25 cc
	12:15	Lunch: $\frac{1}{2}$ soup bowl broth
		1 small bowl Jello
		$\frac{1}{2}$ cup diluted tea
	1:00	Large glass of ginger ale
	1:45	Urine output 250 cc
	1:50	Ice chips—2 cubes
	2:20	IV had absorbed 450 cc
	2:45	Irrigation drainage 100 cc—yellow liquid
	3:00	Patient vomited 200 cc
	3:45	Levine irrigated tube with 15 cc normal saline
	4:00	Supper: 1 soup bowl diluted broth
		1 glass of water
	4:30	Ice chips—5 cubes
	4:45	Ginger ale ($\frac{1}{2}$ juice glass)
	5:25	Urine output of 300 cc
	7:00	Levine tube irrigated with 30 cc normal saline
	7:10	IV had absorbed 350 cc
	7:15	Urine output of 50 cc
	7:40	$\frac{1}{2}$ juice glass of ginger ale
	7:50	Urine output of 100 cc
	9:00	Liquid bowel movement—400 cc yellowish-green liquid
	10:25	1 juice glass of water
	11:15	Urine output of 300 cc
August 15	12:45	$\frac{1}{2}$ glass of water
	1:30	Irrigation drainage of 200 cc clear liquid
	3:45	IV had absorbed 500 cc
	4:00	Levine tube irrigated with 20 cc normal saline
	4:00	Ice chips—3 cubes
	5:30	Urine output of 250 cc
	7:00	Irrigation drainage of 25 cc clear liquid
	7:05	IV had absorbed 350 cc
	7:15	Breakfast: $\frac{1}{2}$ glass of apple juice
		$\frac{1}{2}$ small bowl Jello
		1 large glass of ginger ale
	8:30	Levine tube irrigated with 15 cc normal saline
	9:00	Urine output of 300 cc
	11:10	Ice chips—10 cubes
	12:15	Lunch: 1 large glass of fruit juice
		1 soup bowl broth
		4 oz milk
	1:40	Urine output 250 cc
	1:45	Irrigation drainage—25 cc

Large glass—240 cc   Coffee pot—240 cc
Water glass—180 cc   Small bowl—120 cc
Juice glass—100 cc   Soup bowl—200 cc
Cup—120 cc

Time	Oral	I.V.	Irrig	Remarks	B.M.	Emesis	Urine	Irrig	Remarks
7:00									
8:00									
9:00									
10:00									
11:00									
12:00									
1:00									
2:00									
Shift Total									
3:00									
4:00									
5:00									
6:00									
7:00									
8:00									
9:00									
10:00									
Shift Total									
11:00									
12:00									
1:00									
2:00									
3:00									
4:00									
5:00									
6:00									
Shift Total									
Grand Total									

Large glass—240 cc   Coffee pot—240 cc
Water glass—180 cc   Small bowl—120 cc
Juice glass—100 cc   Soup bowl—200 cc
Cup—120 cc

Time	Oral	I.V.	Irrig	Remarks	B.M.	Emesis	Urine	Irrig	Remarks
7:00									
8:00									
9:00									
10:00									
11:00									
12:00									
1:00									
2:00									
Shift Total									
3:00									
4:00									
5:00									
6:00									
7:00									
8:00									
9:00									
10:00									
Shift Total									
11:00									
12:00									
1:00									
2:00									
3:00									
4:00									
5:00									
6:00									
Shift Total									
Grand Total									

3) Celia Gomez entered Abbey Hospital for colon cancer. She had a section of her colon removed and was placed on intake and output.

September 14 8:30 Breakfast: 1 glass of orange juice
$\frac{1}{2}$ small bowl Jello
2 oz of $\frac{1}{2}$% milk

9:00 IV had absorbed 400 cc. New bag put in place containing 1000 cc.

9:15 Liquid bowel movement—300 cc of dark yellow liquid

9:30 Ice chips—3 cubes

9:40 Levine tube irrigated with 15 cc normal saline

10:05 $\frac{1}{2}$ large glass of ginger ale

10:20 Urine output of 100 cc

10:55 Urine output of 250 cc

11:25 2 oz of pear juice

11:45 Levine irrigated with 20 cc normal saline

12:00 Lunch: 1 large glass of ginger ale
1 small bowl Jello
$\frac{1}{2}$ soup bowl broth

12:30 IV absorption checked. It is now down to 800 cc. (How much has been absorbed?)

12:30 Irrigation drainage with Hemovac—150 cc brown liquid

12:45 Vomited 350 cc yellowish liquid

1:00 Ice chips—5 cubes

1:15 $\frac{1}{2}$ glass of water

1:30 Urine output—250 cc

1:50 Ice chips—2 cubes

2:00 Levine tube irrigated with 30 cc normal saline

2:35 Ice chips—5 cubes

3:00 $\frac{1}{2}$ juice glass of ginger ale

3:20 Urine output—200 cc

3:55 Levine tube irrigated with 15 cc normal saline

4:00 Supper: 1 small bowl Jello
1 cup diluted tea
$\frac{1}{2}$ soup bowl broth

4:35 IV absorption checked. It is now down to 450 cc

5:00 Irrigation drainage 200 cc brownish-yellow liquid

6:30 Urine output—300 cc

8:00 1 large glass ginger ale

8:15 Levine tube irrigated with 20 cc normal saline

8:40 Ice chips—10 cubes

10:15 Urine output—100 cc

11:00 IV bag empty. A new bag put in place containing 1000 cc

September 15 1:45 $\frac{1}{2}$ juice glass of water

4:00 Levine tube irrigated with 20 cc normal saline

4:00 Irrigation drainage of 100 cc yellow liquid

6:30 Urine output of 350 cc

Large glass—240 cc   Coffee pot—240 cc
Water glass—180 cc   Small bowl—120 cc
Juice glass—100 cc   Soup bowl—200 cc
Cup—120 cc

Time	Oral	I.V.	Irrig	Remarks	B.M.	Emesis	Urine	Irrig	Remarks
7:00									
8:00									
9:00									
10:00									
11:00									
12:00									
1:00									
2:00									
Shift Total									
3:00									
4:00									
5:00									
6:00									
7:00									
8:00									
9:00									
10:00									
Shift Total									
11:00									
12:00									
1:00									
2:00									
3:00									
4:00									
5:00									
6:00									
Shift Total									
Grand Total									

Large glass—240 cc   Coffee pot—240 cc
Water glass—180 cc   Small bowl—120 cc
Juice glass—100 cc   Soup bowl—200 cc
Cup—120 cc

Time	Oral	I.V.	Irrig	Remarks	B.M.	Emesis	Urine	Irrig	Remarks
7:00									
8:00									
9:00									
10:00									
11:00									
12:00									
1:00									
2:00									
Shift Total									
3:00									
4:00									
5:00									
6:00									
7:00									
8:00									
9:00									
10:00									
Shift Total									
11:00									
12:00									
1:00									
2:00									
3:00									
4:00									
5:00									
6:00									
Shift Total									
Grand Total									

Chapter 13

Money

Handling Money

Handling money is an essential skill for every kind of medical office. Most medical office personnel will at some time need to make change, collect bill payments, make bank deposits, or sort money.

Money is exchanged in many different forms. It can be currency (bills), coins, checks, money orders, or traveler's checks. Each kind is handled a little differently and should be kept separate in a cash drawer.

Figure 13–1

First, when handling currency, it is helpful to turn all of the bills face up and in one direction as you receive them. Also, keep the largest denominations as far away from the customer's access as possible when placing the money in your cash drawer.

When handling coins, be careful to sort them accurately when placing them in the cash drawer, again keeping the largest denominations as far away from the customer's access as possible.

When handling checks, it is important to go over the check when you receive it to be sure it has been written correctly. Most agencies examine the check for correct address and phone number, correct date, accurate amount (in number and in words), and a complete signature. Other information may be requested and recorded on the back of the check, such as driver's license number, social security number, or credit card numbers.

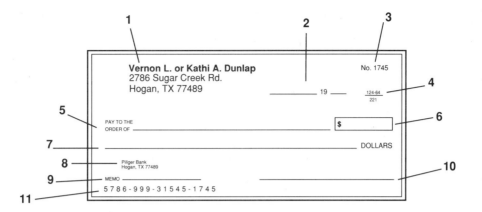

1 Name and address of person(s) holding bank account
2 Date check is written
3 Number of the check
4 Bank identification number
5 Payee of check (person to whom check is written)
6 Amount of check in numbers
7 Amount of check, expressed in words
8 Bank where checking account is held
9 Memo line (reason for writing check)
10 Signature of person writing check
11 Checking account number (last 4 digits indicate check number (item 3) again)

Figure 13–2

In addition, some agencies endorse the check immediately (For Deposit Only—Dr. Augustus Smith) so that it cannot be cashed if it is stolen. Money orders are treated much the same as checks.

Traveler's checks would rarely be seen in a medical office (since most of your patients would have no reason to use traveler's checks in their own town). They can be received, however, and should be signed by the patient in your presence when you receive them. This signature should match the corresponding signature already on the traveler's check.

Making Change

Whenever a patient pays for the services rendered, he may or may not give you the exact payment. If the exact payment is given, accept the money, write the receipt (a procedure introduced in chapter 16), and thank the patient for his prompt payment. If exact payment is not given, accept the money, make correct change, write the receipt, and thank the patient.

Making the correct change is essential. Giving too much change will result in a lack of money in your cash drawer at the end of the day. Giving too little change will create ill will with your patients.

To give change, begin with what the patient owed you and add to that until you reach the amount he or she paid. For example, the bill comes to $27, and the patient gives you two $20 bills ($40 total). Begin at $27 mentally, give him or her three $1 bills ($27 + $3 = $30). Now that you are at $30 mentally, give the patient a $10 bill ($30 + $10 = $40). You have reached $40, so your change is complete. To check, the bill ($27) plus the change ($13) should equal the payment ($40).

BILL + CHANGE GIVEN = PAYMENT

Example 1

The bill is $6.50. The patient gives you a $20 bill.

		Say...	
		6.50	(begin mentally)
give	.25	6.75	(trying to get to an even dollar amount)
give	.25	7.00	(now use $1 bills)
give	1.00	8.00	
give	1.00	9.00	(trying to get to an even $5 or $10)
give	1.00	10.00	(now use $5, $10, or $20 bills)
give	<u>10.00</u>	20.00	
	13.50		

Check: 6.50 (bill) + 13.50 (change) = $20 (payment)

Example 2

The bill is $4.79. The patient gives you a $10 bill.

		Say...	
		4.79	(begin mentally)
give	.01	4.80	(now you can use dimes)
give	.10	4.90	(trying to get to an even dollar amount)
give	.10	5.00	(now use $5, $10, or $20 bills)
give	<u>5.00</u>	10.00	
	5.21		

Check: 4.79 (bill) + 5.21 (change) = $10 (payment)

Problems   Use only the spaces needed to complete each problem.

1) The bill is $29. The patient gives you two $20 bills.

    Say...

    _____

    give _____ _____

    give _____ _____

    give _____ _____

    give _____ _____

2) The bill is $3.50. The patient gives you a $20 bill.

    Say...

    _____

    give _____ _____

    give _____ _____

    give _____ _____

    give _____ _____

    give _____ _____

    give _____ _____

3) The bill is $6.95. The patient gives you a $10 bill.

Say…

_____

give   _____   _____
give   _____   _____
give   _____   _____
give   _____   _____
give   _____   _____
give   _____   _____

4) The bill is $18. The patient gives you a $50 bill.

Say…

_____

give   _____   _____
give   _____   _____
give   _____   _____
give   _____   _____
give   _____   _____

5) The bill is $2.88. The patient gives you a $10 bill.

Say…

_____

give   _____   _____
give   _____   _____
give   _____   _____
give   _____   _____
give   _____   _____
give   _____   _____

6) The bill is $26.69. The patient gives you a $20 bill and a $10 bill.

Say…

_____

give   _____   _____
give   _____   _____
give   _____   _____
give   _____   _____
give   _____   _____
give   _____   _____
give   _____   _____

7) The bill is $86.75. The patient gives you five $20 bills.

Say...

_____

give _____ _____
give _____ _____
give _____ _____
give _____ _____
give _____ _____
give _____ _____

8) The bill is $14.60. The patient gives you a $10 bill and a $5 bill.

Say...

_____

give _____ _____
give _____ _____
give _____ _____
give _____ _____
give _____ _____
give _____ _____

9) The bill is $39.00. The patient gives you two $20 bills.

Say...

_____

give _____ _____
give _____ _____
give _____ _____
give _____ _____
give _____ _____

10) The bill is $106.15. The patient gives you six $20 bills.

Say...

_____

give _____ _____
give _____ _____
give _____ _____
give _____ _____
give _____ _____
give _____ _____
give _____ _____
give _____ _____

In problems 11–20, write in order what coins or bills you would give.

Example: Bill: $14.55
         Given: $20.00

| $.10 | $.10 | $.25 | $5.00 | ___ | ___ | ___ | ___ | ___ | ___ | ___ |

11) Bill: $21.05
     Given: $25.00

| ___ | ___ | ___ | ___ | ___ | ___ | ___ | ___ | ___ | ___ | ___ |

12) Bill: $13.73
     Given: $20.00

| ___ | ___ | ___ | ___ | ___ | ___ | ___ | ___ | ___ | ___ | ___ |

13) Bill: $47.09
     Given: $50.00

| ___ | ___ | ___ | ___ | ___ | ___ | ___ | ___ | ___ | ___ | ___ |

14) Bill: $116.48
     Given: $120.00

| ___ | ___ | ___ | ___ | ___ | ___ | ___ | ___ | ___ | ___ | ___ |

15) Bill: $18.80
     Given: $50.00

| ___ | ___ | ___ | ___ | ___ | ___ | ___ | ___ | ___ | ___ | ___ |

16) Bill: $43.75
     Given: $60.00

| ___ | ___ | ___ | ___ | ___ | ___ | ___ | ___ | ___ | ___ | ___ |

17) Bill: $9.00
     Given: $20.00

| ___ | ___ | ___ | ___ | ___ | ___ | ___ | ___ | ___ | ___ | ___ |

18) Bill: $57.40
     Given: $100.00

| ___ | ___ | ___ | ___ | ___ | ___ | ___ | ___ | ___ | ___ | ___ |

19) Bill: $23.95
     Given: $30.00

| ___ | ___ | ___ | ___ | ___ | ___ | ___ | ___ | ___ | ___ | ___ |

20) Bill: $74.20
     Given: $75.00

| ___ | ___ | ___ | ___ | ___ | ___ | ___ | ___ | ___ | ___ | ___ |

Collecting Money

All medical offices should have a collection policy established. This is a procedure explaining the office's policy for payment of bills. This policy usually includes how to ask for payment from customers, how often billing will occur, and when accounts will be handed over to a collection agency.

> **Payment is expected when services are rendered. This helps reduce billing costs. Your cooperation is appreciated.**
>
> **Dr. McGuire & Staff**

Figure 13–3

The policy is made known to the patients by a sign at the desk such as, All Charges Should be Paid at the Time of Service, a letter sent from the office, a patient information booklet, or by the doctor and his staff. It is the duty of the staff to be firm in requesting collection or making billing arrangements.

Any time a bill is not paid in full, a payment plan should be set up to encourage payment of the remaining money. When patients first visit the office, it is wise to have them complete a detailed form called a credit history.

A credit history should include all of the following information:

Patient's full name, address, and phone number

Spouse's name

Date of birth, social security number, driver's license number

Name, address, and phone number of person responsible for payment

Name of landlord or mortgage holder

Bank affiliations

Place of employment for patient, spouse, and/or responsible person

Names, addresses, and phone numbers of two other relatives not living with patient

After the patient has completed this form (see figure 13–4) and is accepted as a patient in your office, you are responsible for being firm in requesting money from him. One of the best ways of discussing the patient's bill is the effective use of silence. For example, you might say, "Mr. Smith, the charge today is thirty dollars. Will you be paying by cash or check?" Then be silent. This pause will encourage the patient to respond.

Often patients do not pay for services because no one has ever asked. It is a good policy, though, to expect payment at the time of service. If a patient cannot pay the total amount, you should ask, "How much will you pay today?" Then be silent. Or a patient may ask for a bill to be sent. In this case, explain that sending bills increases office costs and request at least partial payment. For example, "We are trying to keep our costs down so we can keep our fees reasonable. Could you make partial payment today?"

```
┌──────────────────────────────────────────────────────────┐
│  Name of Insured_____  Phone _____  │
│                                                            │
│  Address_____  Zip Code_____    │
│                                                            │
│  Date of Birth _____ Social Security No._____  │
│                                                            │
│  Employer _____ Work Phone_____   │
│                                                            │
│  Address _____ Zip Code _____     │
│                                                            │
│  Insurance Carrier _____   │
│                                                            │
│  Address_____    │
│                                                            │
│  Identification No./Subscriber No. _____    │
│                                                            │
│                                                            │
│  Insured Dependent Information                             │
│        Name              Relationship        Birthdate     │
│  Spouse _____    │
│                                                            │
│  Others _____    │
│         _____    │
│         _____    │
│         _____    │
│         _____    │
│         _____    │
│                                                            │
│  Additional Coverage  YES/NO  If so, Carrier_____    │
│                                                            │
│  Address_____    │
│                                                            │
│  Identification No./Subscriber No. _____    │
│                                                            │
└──────────────────────────────────────────────────────────┘
```

Figure 13–4

When accounts are not paid in full, it is useful to age the accounts. Account aging is a list or table of who owes how much money from services during the past month, those with accounts two months old, those with accounts three months old, and finally those with accounts that are four or more months old.

There are three ways to complete an account aging list. The first is to use a computerized system. The next easiest way is to identify the age of your accounts by color coding them (yellow stickers for one-month-old accounts, blue stickers for two months, etc.). The third way is to make an actual list at the end of every month. This, however, can be very time consuming.

Problems   Answer the following questions.

1) Define account aging. _____

   _____

   _____

2) What are three ways of compiling an aging list?

   a. _____

   b. _____

   c. _____

3) List the information that a credit history should include:

   a. _____

   b. _____

   c. _____

   d. _____

e. _____

f. _____

g. _____

h. _____

4) How might the collection policy of an office be made known to your patients?

a. _____

b. _____

c. _____

d. _____

5) What is one reason a patient does not pay for services?

_____

_____

Balancing a Cash Drawer

When all money has been received for the day, a bank deposit must be prepared. Before the money can be deposited, however, a cashier count sheet must be completed to determine if the amount collected from patients is correct (figure 13–5).

Daily Balance Sheet

Date: _____

NUMBER	DENOMINATION	AMOUNT
	Pennies	
	Nickels	
	Dimes	
	Quarters	
	Half Dollars	
	$1 Bills	
	$5 Bills	
	$10 Bills	
	$20 Bills	
	Checks	
	Cash in Drawer	
	Plus Cash Paid Out	
	Total Cash	
	Less Change	
	Cash Received, Cashier's Count	
	Cash Received, Ledger Count	
	AMOUNT OF CASH SHORT OR OVER	

Cash Proved _____   Cash Over _____   Cash Short _____

Figure 13–5

To complete the cashier count sheet, first separate checks, money orders, traveler's checks, and cash. You should then count how many of each different kind of coin and bill you have. This information is recorded in the left hand column under **number**. When this has been done (and double checked), set the money aside. Next, multiply the number of coins or bills by the value of that coin or bill. This gives you the **amount** to be recorded in the right-hand column. The checks, money orders, and traveler's checks should be counted next, then the value of these items should be totaled. Set checks, money orders, and traveler's checks aside.

After this count, all amounts should be totaled to determine **cash in drawer**.

Any **cash paid outs** (money paid from the cash drawer for deliveries, etc.) should be written on the next line and added to the cash in drawer to determine **total cash**.

Next, write the amount of cash you began the day with (petty cash for making change) in **less change** and subtract it from the total cash to determine **cash received, cashier's count**.

This is the amount that you, as cashier, have collected that day. After this is complete, write in the amount of collection that your cash register or day sheet indicates was collected that day on the **cash received, ledger** line. Then compare your cashier count with the ledger amount. If the cashier count is the smaller amount, you are short of funds. If the cashier count is larger, you are over. If the amount is the same, it is said that your cash "proved." Identify whether you are short, over, or proved, and check the appropriate line. Then subtract the two amounts to find the difference. Write this amount at the bottom of the form in **amount of cash short or over**.

Problems   Complete these cashier count sheets using the following information:

1)
pennies	37
nickels	16
dimes	30
quarters	19
half dollars	0
$1 bills	41
$5 bills	7
$10 bills	5

$20 bills	6
$50 bills	0
checks	5 totaling $147.89
cash paid out	$ 4.25
beginning change	40.00
cash received according to ledger sheet or cash register	376.06

2)
pennies	26
nickels	15
dimes	17
quarters	21
half dollars	0
$1 bills	18
$5 bills	9
$10 bills	8

$20 bills	5
$50 bills	0
checks	5 totaling $189.10
cash paid out	$ 7.15
beginning change	25.00
cash received according to ledger sheet or cash register	422.21

3)
pennies	20
nickels	17
dimes	12
quarters	19
half dollars	0
$1 bills	27
$5 bills	16
$10 bills	14

$20 bills	7
$50 bills	0
checks	6 totaling $210.00
cash paid out	$ 1.30
beginning change	40.00
cash received according to ledger sheet or cash register	570.30

Daily Balance Sheet

Date: _____

NUMBER	DENOMINATION	AMOUNT
	Pennies	
	Nickels	
	Dimes	
	Quarters	
	Half Dollars	
	$1 Bills	
	$5 Bills	
	$10 Bills	
	$20 Bills	
	Checks	
	Cash in Drawer	
	Plus Cash Paid Out	
	Total Cash	
	Less Change	
	Cash Received, Cashier's Count	
	Cash Received, Ledger Count	
	AMOUNT OF CASH SHORT OR OVER	

Cash Proved _____   Cash Over _____   Cash Short _____

Daily Balance Sheet

Date: _____

NUMBER	DENOMINATION	AMOUNT
	Pennies	
	Nickels	
	Dimes	
	Quarters	
	Half Dollars	
	$1 Bills	
	$5 Bills	
	$10 Bills	
	$20 Bills	
	Checks	
	Cash in Drawer	
	Plus Cash Paid Out	
	Total Cash	
	Less Change	
	Cash Received, Cashier's Count	
	Cash Received, Ledger Count	
	AMOUNT OF CASH SHORT OR OVER	

Cash Proved _____   Cash Over _____   Cash Short _____

Daily Balance Sheet

Date: _____

NUMBER	DENOMINATION	AMOUNT
	Pennies	
	Nickels	
	Dimes	
	Quarters	
	Half Dollars	
	$1 Bills	
	$5 Bills	
	$10 Bills	
	$20 Bills	
	Checks	
	Cash in Drawer	
	Plus Cash Paid Out	
	Total Cash	
	Less Change	
	Cash Received, Cashier's Count	
	Cash Received, Ledger Count	
	AMOUNT OF CASH SHORT OR OVER	

Cash Proved _____   Cash Over _____   Cash Short _____

4) pennies 13
    nickels 19
    dimes 21
    quarters 40
    half dollars 0
    $1 bills 17
    $5 bills 8
    $10 bills 6

    $20 bills 5
    $50 bills 0
    checks .... 5 totaling $174.00
    cash paid out .... $ 7.70
    beginning change 20.00
    cash received according to ledger
      sheet or cash register 390.88

5) pennies 9
    nickels 12
    dimes 11
    quarters 17
    half dollars 0
    $1 bills 25
    $5 bills 10
    $10 bills 11

    $20 bills 7
    $50 bills 0
    checks .... 4 totaling $110.50
    cash paid out .... $ 6.00
    beginning change 35.00
    cash received according to ledger
      sheet or cash register 302.04

Daily Balance Sheet

Date:_____

NUMBER	DENOMINATION	AMOUNT
	Pennies	
	Nickels	
	Dimes	
	Quarters	
	Half Dollars	
	$1 Bills	
	$5 Bills	
	$10 Bills	
	$20 Bills	
	Checks	
	Cash in Drawer	
	Plus Cash Paid Out	
	Total Cash	
	Less Change	
	Cash Received, Cashier's Count	
	Cash Received, Ledger Count	
	AMOUNT OF CASH SHORT OR OVER	

Cash Proved _____   Cash Over _____   Cash Short _____

Daily Balance Sheet

Date:_____

NUMBER	DENOMINATION	AMOUNT
	Pennies	
	Nickels	
	Dimes	
	Quarters	
	Half Dollars	
	$1 Bills	
	$5 Bills	
	$10 Bills	
	$20 Bills	
	Checks	
	Cash in Drawer	
	Plus Cash Paid Out	
	Total Cash	
	Less Change	
	Cash Received, Cashier's Count	
	Cash Received, Ledger Count	
	AMOUNT OF CASH SHORT OR OVER	

Cash Proved _____   Cash Over _____   Cash Short _____

Chapter 14

Time

Elapsed Time

Elapsed time is the amount of time that has passed from one moment to another. We often use elapsed time without thinking much about it. For example, in a doctor's office it may be used to determine when to schedule the next appointment or how long to leave a thermometer in place. These are common uses of elapsed time that can often be figured mentally. Another use of elapsed time in nearly every medical office is determining how long an employee has been at work. This information is needed to compute the employee's paycheck.

To figure elapsed time, there are a few basic rules:

1. Put the ending (or leaving) time on top of the starting (or arrival) time as for a subtraction problem. Separate the hours from the minutes with a vertical line.

2. Subtract the minutes. If you are unable to and need to borrow, borrow one hour from the hours and add that hour (as 60 minutes now) to the minutes. Then subtract.

3. Next, subtract the hours. If you are unable to, add 12 hours (a full rotation around the clock). Then subtract.

Example 1:

Find the elapsed time from 9:11 am to 3:15 pm.

1)
```
   3: | 15
 - 9: | 11
```
Set up subtraction and draw a vertical line (note: This is *not* 315 minus 911).

2)
```
   3: | 15
 - 9: | 11
 ─────────
      | 04
```
Subtract minutes (15 − 11 = 4). Write as a two-digit answer (04).

3)
```
   15
   3̸: | 15
 - 9: | 11
 ─────────
   6: | 04
```
Subtract hours (3 − 9 = impossible). If unable to, add 12 hours to 3, then subtract.

4) Elapsed time is 6 hours and 4 minutes. This may be written as 6:04, but remember that it does not mean 6:04 am or pm; it means a time of 6 hours and 4 minutes elapsed.

Example 2:

Find the elapsed time from 1:14 pm to 9:29 pm.

1)
```
   9: | 29
 - 1: | 14
```
Set up subtraction and draw a vertical line.

2)   9: | 29   Subtract minutes (29 − 14 = 15). Write as a two-digit answer (15).
       − 1: | 14
            15

3)   9: | 29   Subtract hours (9 − 1 = 8).
       − 1: | 14
       8: 15

4) Elapsed time is 8 hours and 15 minutes.

Example 3:

Find the elapsed time from 11:58 am to 5:05 pm.

1)   5: | 05   Set up subtraction and draw a vertical line.
     − 11: | 58

2)   4   ⁶⁵   Subtract minutes (5 − 58 = impossible). If unable to, borrow an hour (so 5 hours
     5̸: | 0̸5   becomes 4) and add that hour (as 60 minutes) to the 5 minutes (60 + 5 = 65), then
   − 11: | 58   subtract.
         07

3) **16**   Subtract hours (4 − 11 = impossible). If unable to, add 12 hours to 4, then subtract.
   4̸   ⁶⁵
    5̸: | 0̸5
   − 11: | 58
    5: 07

4) Elapsed time is 5 hours and 7 minutes.

Often when recording time for time sheets, employees will record only the time they are actually at work, excluding their lunch hour. To solve a problem like this, compute elapsed time for the morning hours and then the evening hours and add them together. Be aware when adding minutes that any total which is larger than 59 minutes is not written as 60 minutes or more but as 1 hour or 1 hour and so many minutes instead.

Example 1:

Find the total elapsed time from 8:07 am to 11:45 am and from 12:30 pm to 5:06 pm.

1) Morning hours   Afternoon hours

                                            16
                                        4̸   ⁶⁶
     11: | 45   5̸: | 0̸6
     − 8: | 07   − 12: | 30
      3: | 38   4: | 36

2) Add   3: | 38   74 minutes is 1 hour, 14 minutes. Add the
           + 4: | 36   1 hour to the 7, and deduct 60 minutes
             7: | 74   from the 74.

             8: | 14

3) Solution is 8:14, or 8 hours and 14 minutes.

Example 2:

Find the total elapsed time from 7:09 am to 12:56 pm and from 2:01 pm to 5:58 pm.

1) Morning hours   Afternoon hours

$$
\begin{array}{r|r}
12: & 56 \\
-\ 7: & 09 \\
\hline
5: & 47 \\
\end{array}
\qquad\qquad
\begin{array}{r|r}
5: & 58 \\
-\ 2: & 01 \\
\hline
3: & 57 \\
\end{array}
$$

2) Add $\begin{array}{r|r} 5: & 47 \\ +\ 3: & 57 \\ \hline 8: & 104 \\ 9: & 44 \\ \end{array}$ 104 minutes is 1 hour 44 minutes. Add the 1 hour to the 8, and deduct 60 minutes from the 104.

3) Solution is 9:44, or 9 hours and 44 minutes.

Problems   Find the elapsed time.

1) 9:18 am to 3:15 pm   6) 9:02 am to 5:11 pm

2) 10:00 am to 5:48 pm   7) 6:49 am to 12:08 pm

3) 8:19 am to 3:30 pm   8) 12:43 pm to 5:06 pm

4) 9:00 am to 3:57 pm   9) 11:40 am to 6:07 pm

5) 7:58 am to 4:14 pm   10) 7:45 am to 3:44 pm

In problems 11–15, find elapsed time for am and pm and add together.

	IN	OUT	IN	OUT	
11)	7:58	11:49	1:00	4:56	am _____
					pm _____
					Total _____
12)	8:01	12:51	1:30	4:19	am _____
					pm _____
					Total _____
13)	6:58	11:02	12:00	4:00	am _____
					pm _____
					Total _____
14)	7:00	11:30	12:15	4:30	am _____
					pm _____
					Total _____
15)	8:06	11:53	12:31	4:08	am _____
					pm _____
					Total _____

Time Sheets

In a medical office, each employee is required to keep a written record of the time he or she is present at work. This written record is called a **time sheet**. The office manager or payroll clerk uses the time sheet to compute each employee's paycheck.

Each office has different policies concerning lateness and/or overtime. For example, one office might deduct earnings for every minute you are late, while other offices have a five or seven minute "grace period." The same is true with regular hours worked and overtime. Some offices pay overtime for any minute over, while others will not pay overtime unless 15 or more minutes overtime have been worked.

Another system that might be used is to total the hours worked each day to the exact minute, then round that to the nearest quarter of an hour. This is the system we will use to compute time sheets. To find the nearest quarter hour, determine where the minutes are on the clock and decide to which quarter hour it is closest (:00, :15, :30, :45).

Example 1:

8:12 is closer to 8:15 than it is to 8:00.

Example 2:

8:22 is closer to 8:15 (7 minutes away) than it is to 8:30 (8 minutes away).

When the hours worked each day have been rounded to the nearest quarter hour, the entire week is then added together.

Example 1:

Suppose you total each day of the week and get the following results. First, round each day to the nearest quarter hour and then total the week.

	Total	*Rounded*
Monday	8:41	8:45
Tuesday	7:56	8:00
Wednesday	8:05	8:00
Thursday	9:46	9:45
Friday	7:22	7:15
		40:105 (do not carry)

105 minutes = 1 hr and 45 minutes
40:105 becomes 41:45
Final answer: 41:45, or 41 hours and 45 minutes.

Problems Round each day to the nearest quarter hour, then total the week.

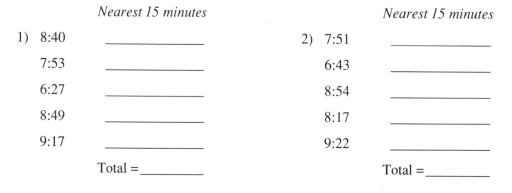

		Nearest 15 minutes			*Nearest 15 minutes*
1)	8:40	_____	2)	7:51	_____
	7:53	_____		6:43	_____
	6:27	_____		8:54	_____
	8:49	_____		8:17	_____
	9:17	_____		9:22	_____
	Total =	_____		Total =	_____

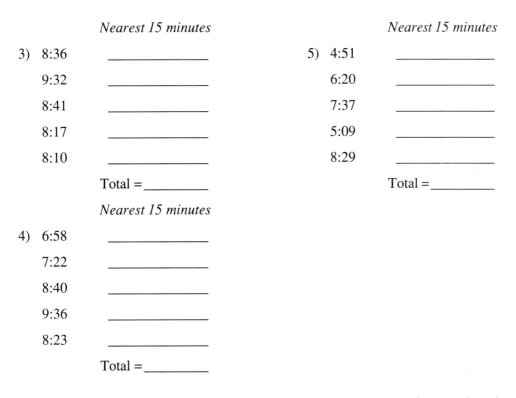

Nearest 15 minutes

3) 8:36   _____
    9:32   _____
    8:41   _____
    8:17   _____
    8:10   _____
            Total = _____

Nearest 15 minutes

5) 4:51   _____
    6:20   _____
    7:37   _____
    5:09   _____
    8:29   _____
            Total = _____

Nearest 15 minutes

4) 6:58   _____
    7:22   _____
    8:40   _____
    9:36   _____
    8:23   _____
            Total = _____

6) Figure the elapsed time for each day, round to the nearest quarter hour, and total.

	IN	OUT	IN	OUT	TOTAL	NEAREST 15 MIN.
Mon.	7:48	11:22	12:08	5:00	am _____	
					pm _____	
					Total _____	_____
Tue.	8:11	11:46	12:30	4:19	am _____	
					pm _____	
					Total _____	_____
Wed.	9:00	11:30	12:05	5:00	am _____	
					pm _____	
					Total _____	_____
Thurs.	7:45	12:08	12:45	4:30	am _____	
					pm _____	
					Total _____	_____
Fri.	8:00	1:18	2:00	5:36	am _____	
					pm _____	
					Total _____	_____

Chapter 15

Income

Gross Pay

Gross pay is the amount of pay earned for a given period of time. Common time periods used for payment are weekly, monthly, biweekly, and semimonthly. Those who are paid weekly receive a check 52 times a year. Those who are paid monthly receive a check 12 times a year. Biweekly means every two weeks (or every other week on the same day of the week, e.g., every other Friday) and therefore occurs 26 times a year. Semimonthly means every half a month (twice a month on designated dates, e.g., the first and fifteenth of each month) and therefore occurs 24 times a year.

Figure 15–1

The type of job you have determines how your pay will be computed. In the medical field, pay is most often computed by the hour (with straight time and overtime) or by salary.

Hourly Pay

To figure straight time, multiply the regular hours worked by the pay received per hour.

Example 1

June worked 37 hours and earns $4.29 per hour. Compute her gross pay.

$$\$4.29 \times 37 = \$158.73$$

Overtime pay is either computed at time and a half (one and a half times as much money per hour) or double time (twice as much per hour). To figure overtime pay, multiply the pay received per hour by either 1.5 for time and a half or 2 for double time. Then, multiply that amount per hour by the number of overtime hours worked.

Example 2

Randall worked 40 regular hours and 6 hours overtime. His company pays overtime at time and a half. He earns $6.92 per hour. Find his straight-time pay, his overtime pay, and his total pay.

Straight time	**Overtime**
$6.92	$6.92
× 40 hours	× 1.5 time and a half
$276.80	$10.38 per hour for overtime
	× 6 hours overtime
	$62.28

$276.80 + $62.28 = $339.08 total pay

Salary

To figure gross pay for an individual who earns a salary (a fixed amount per year regardless of hours worked), divide the salary amount by how many pay periods are in one year. For example, if the worker is paid weekly, you would divide by 52; biweekly, 26; monthly, 12; and semimonthly, 24.

Example 1

Jack Reams earns $27,590 per year as a pharmacist. He is paid biweekly. What is his gross pay?

biweekly = 26

$$26\overline{)27,590.}\quad 1,061.153$$

Use thousandths place to decide how to round.

Gross pay is $1,061.15

Problems   Determine each individual's gross pay.

1) RaeAnn works as a receptionist and earns $5.71 per hour. What is her gross pay for a week in which she worked 38.5 hours?   _____

2) Dean earns $12.89 per hour as a sports medicine technician. What is his gross pay for a week in which he worked 40 hours?   _____

3) Justine works $37\frac{3}{4}$ hours a week as a veterinarian's assistant. She earns $8.10 per hour. What is her gross pay for the week?   _____

4) Jeff works at Davidson Company Health Center as a counselor. What is his gross pay for a week in which he works 25 hours and earns $6.45 per hour?   _____

5) Laura Montgomery works as a public health consultant. She earns $11.21 per hour and earns time and a half for overtime. What is her gross pay for a week in which she works 40 regular hours and $6\frac{1}{2}$ hours overtime?   _____

6) Rick earns $5.37 per hour as a nurse's assistant at Montpelier Home Center. What is his gross pay in a week in which he works $41\frac{1}{4}$ regular hours?   _____

7) Carol works at Dr. Kline's office as a dental assistant. She earns $6.40 per hour and gets double time for overtime. What is her pay for the week in which she works 40 regular hours and $7\frac{1}{2}$ overtime hours?   _____

8) Darla earns $11.90 per hour as a physical therapist. She worked 30 regular hours last week. What was her gross pay?   _____

9) Shannon works at Medical Supplies, Inc., as a researcher. She earns $10.88 per hour and double time for overtime. What is her pay for a week in which she works 40 regular hours and 7 overtime hours?   _____

10) Leslie Wireman works for Heart Ambulance as a dispatcher. She is paid $7.01 per hour and earns time and a half for overtime on Saturday and double time for overtime on Sunday. What is her pay for a week in which she worked 40 regular hours, 6 hours on Saturday, and 4 hours on Sunday?   _____

11) Joan earns $14,975 a year as a nutrition counselor for the elderly. What is her gross pay each week?   _____

12) Scott earns $21,807 a year as a registered nurse. What is his biweekly gross pay?   _____

13) Wayne gets paid once a month in his job as a veterinarian. What is his monthly gross pay if he earns $36,496 a year?   _____

14) Jill earns $20,411 a year as a school nurse. What is her semimonthly pay?   _____

15) Rick is an orderly at Bay View Hospital. He earns $4.90 per hour. What is his weekly pay for a week in which he works $37\frac{1}{2}$ regular hours?   _____

16) Match the following pay periods with the correct description:

Pay Day(s)	Description	Number of Periods
A. the 5th of every month	_____ biweekly	_____ 24
B. every Tuesday	_____ monthly	_____ 52
C. the 10th and 25th of every month	_____ weekly	_____ 12
D. every other Friday	_____ semimonthly	_____ 26

Deductions

Federal Income Tax

Federal Income Tax (FIT) is deducted from each paycheck based on your marital status, your income, and how many allowances you claim.

An allowance is a credit for each person you financially support (yourself, your spouse, children, older parents, etc.). You may claim as many people as you support, but you are not required to claim all of them.

If you claim all allowances, less will be deducted from each paycheck. However, you probably will not get a very large tax refund at the end of the year, and you may even have to pay.

If you claim fewer allowances, more will be deducted from each paycheck. This will probably "store up" enough with the government that you will get a larger refund. Some individuals see this as a way to save money for a special purpose.

The amount deducted from an individual's paycheck is determined by reading a table. There are separate tables for singles and for those who are married. A single person has more deducted per week than a married person with the same income and allowances.

To determine FIT withheld:
1. Choose the appropriate table for marital status.
2. Look for the row that includes the income amount (the amount *must* be weekly income).
3. Read over to the appropriate number of allowances.
4. The amount from the table is the FIT withheld.

Problems Find each individual's FIT withheld.

		Marital status	Allowances	Weekly income	FIT withheld
1)	Joe Belaquin	M	2	$211.80	_____
2)	Jane Milligan	M	1	$388.16	_____
3)	Laura Hillshire	S	0	$270.00	_____
4)	Sarina Green	M	4	$241.87	_____
5)	Rico Gomez	S	0	$349.99	_____
6)	Allen Silvers	M	3	$353.44	_____
7)	Carla Pennington	S	1	$274.16	_____
8)	Bob Dome	S	1	$202.08	_____
9)	JaShonda Michaels	S	0	$330.00	_____
10)	Ray Everett	M	4	$376.55	_____

Table 15–1   Weekly Payroll Deduction for FIT for Married Persons

At least	But less than	0	1	2	3	4	5	6
200	210	21.08	18.12	15.07	12.15	9.07	6.17	3.07
210	220	22.04	19.06	16.03	13.08	10.06	7.02	4.06
220	230	22.97	20.01	16.95	14.08	10.98	8.05	4.98
230	240	23.88	20.75	17.88	14.74	11.83	8.79	5.83
240	250	24.75	21.01	18.75	15.05	12.79	9.04	6.79
250	260	25.61	22.17	19.62	16.12	13.64	10.14	7.64
260	270	26.82	23.64	20.83	17.68	14.85	11.67	8.85
270	280	27.50	24.25	21.58	18.23	15.52	12.22	9.52
280	290	29.00	25.73	23.05	19.76	17.01	13.78	11.01
290	300	30.63	26.99	24.68	20.91	18.64	14.94	12.64
300	310	31.98	27.50	25.93	21.58	19.98	15.50	13.98
310	320	33.14	29.95	27.16	23.93	21.19	17.92	15.19
320	330	34.78	30.78	28.77	24.70	22.75	18.75	16.75
330	340	36.22	32.77	30.23	26.76	24.26	20.76	18.26
340	350	37.76	34.54	31.77	28.54	25.73	22.59	19.73
350	360	39.05	35.17	33.09	29.18	27.09	23.14	21.09
360	370	40.44	37.80	34.47	31.83	28.44	25.89	22.44
370	380	41.70	38.27	35.79	32.27	29.70	26.24	23.70
380	390	42.93	39.00	36.95	33.04	30.90	27.07	24.90
390	400	44.21	40.85	38.26	34.89	32.25	28.89	26.25

The wages are: / The withholding allowances are

Table 15–2   Weekly Payroll Deduction for FIT for Single Persons

At least	But less than	0	1	2	3	4	5	6
200	210	24.04	21.14	18.07	15.14	12.03	9.12	6.05
210	220	25.04	22.08	19.03	16.07	13.04	10.03	7.08
220	230	25.97	23.05	19.95	17.05	13.97	11.06	7.93
230	240	26.84	23.79	20.88	17.79	14.82	11.78	8.88
240	250	27.72	24.02	21.75	18.03	15.79	12.04	9.74
250	260	28.67	25.14	22.62	19.15	16.60	13.19	10.65
260	270	29.84	26.65	23.83	20.62	17.86	14.60	11.88
270	280	30.59	27.27	24.58	21.21	18.59	15.24	12.53
280	290	31.05	28.70	26.05	22.74	20.08	16.78	14.05
290	300	33.64	29.97	27.68	23.98	21.63	17.92	15.68
300	310	34.98	30.59	28.93	24.50	22.95	18.52	16.99
310	320	36.12	32.94	30.16	26.95	24.12	20.96	18.12
320	330	37.79	33.75	31.77	27.76	25.71	21.71	19.75
330	340	39.24	35.77	33.23	29.79	26.27	23.77	21.22
340	350	40.74	37.53	34.77	31.53	28.74	25.55	22.79
350	360	42.05	38.19	36.09	32.19	30.09	26.19	24.00
360	370	43.48	39.86	37.47	34.85	31.40	28.83	25.44
370	380	44.72	41.29	38.79	35.28	32.77	29.29	26.76
380	390	45.99	42.08	39.95	36.04	33.99	30.04	27.92
390	400	47.20	43.84	41.26	37.89	35.24	31.88	29.28

The wages are: / The withholding allowances are

11) Coretta Shields earns $271.49 a week as a lab technician. She is divorced and claims 3 allowances. What is her FIT withheld? _____

12) Troy Blake is married and supports his wife and three children. He claims all of the allowances he is entitled to claim.

   a. How many allowances does he claim? _____

   b. How much FIT is withheld if he earns $380.00 per week? _____

13) Jennifer Carpenter recently got married. When she was single, she claimed 1 allowance. Now that she is married, she will claim 0. Her weekly income is $245.48.

   a. What was the amount of FIT withheld when she was single? _____

   b. What is her FIT withheld now that she is married? _____

14) Duane Shearer earns $18,200 a year as an office manager at Moland Dental Care. He is married and claims 3 allowances. What is his FIT withheld? _____

15) Vanessa Beauvé operates a child-care center. She earns $20,277.40 a year. She is single and claims 0 allowances. What is her FIT withheld? _____

State Income Tax

State Income Tax is deducted from each paycheck based on your marital status, your income, and how many dependents you have.

For each person you support, you receive an exemption on your state taxes. An exemption is a credit for each person you financially support (yourself, your spouse, children, older parents, etc.). The table below shows an example of the exemptions a state might allow.

Personal Exemptions	
Single	$2,000
Married	4,000
Each dependent	1,000

This amount is deducted from your annual income to determine your taxable income. Taxable income is then multiplied by the state tax rate to determine annual state tax.

Example 1

Jeff Reed earns $14,500 a year as an X-ray technician. He is married and has 1 dependent. His state has a $2\frac{1}{2}\%$ tax rate. Find the annual state tax.

1. Find the taxable income (annual income – exemptions).

$$\$14,500 - \underset{\text{married}}{4,000} - \underset{\text{dependent}}{1,000} = \$9,500$$

2. Find annual state tax (taxable income × rate).

$$\$9,500 \times 2.5\% = \$237.50$$

Problems   Find each individual's state income tax withheld. Use the table provided for exemptions.

		State Tax Rate	Marital Status	Dependents	Annual Income	Annual State Tax
1)	D. Bryan	2.5%	M	2	$7,900	_____
2)	S. Evans	4.1%	S	0	11,250	_____
3)	C. Carr	3.75%	S	1	14,115	_____
4)	S. Hale	5%	M	4	26,630	_____
5)	R. Callas	4.25%	S	1	20,050	_____
6)	P. Reese	3%	M	3	16,800	_____
7)	J. Cole	3.2%	M	1	18,000	_____
8)	L. Wheeler	4.25%	S	0	19,135	_____
9)	J. Juarez	4%	M	2	15,885	_____
10)	A. Blair	2%	S	1	11,488	_____

11) Sheree Cabon works at Montpelier Medical Supplies as a sales representative. She earns $361.00 a week. She is married and has 2 dependents. Her state has a $1\frac{1}{2}$% tax rate.

   a. Find her annual salary.   _____

   b. Find her annual state tax withheld.   _____

12) Joy Petrie earns $7.22 per hour as an admission clerk at Rois Hospital. She works 40 hours a week, 52 weeks a year. She is single and has 0 dependents. Her state tax rate is 4.24%.

   a. Find her annual salary.   _____

   b. Find her annual state tax withheld.   _____

13) Angela Windela earns $18,788 a year as an office manager at Dillon Dentistry. She is single and has no dependents. Her state tax rate is 2.75%.

   a. Find her annual state tax withheld.   _____

   b. Find the amount withheld from each paycheck if she is paid semimonthly.   _____

14) Ray Bolson earns $189.88 weekly as a dental assistant. He is single and has 1 dependent. His state tax rate is 4%.

   a. Find his annual salary.   _____

   b. Find his annual state tax withheld.   _____

   c. Find his weekly state tax withheld.   _____

15) Roger Waycliffe is an RN at Phillips Memorial Hospital. He is married and has 4 dependents. He earns $1715.89 monthly. His state tax rate is 6.2%.

   a. Find his annual salary.   _____

   b. Find his annual state tax withheld.   _____

   c. Find his weekly state tax withheld.   _____

FICA Tax (Social Security Tax)

   Federal Insurance Contributions Act (FICA) tax is deducted from each paycheck at a rate of 7.51%, up to a total earning of $45,000 a year. After $45,000 has been earned by the individual, he or she no longer pays social security tax that year; he or she begins paying it again at the beginning of the next year.

To compute FICA tax, it is important to first determine whether the individual has earned over $45,000 so far during the year. If so, FICA tax withheld will be zero for that pay period and all others remaining that year.

If they have not yet earned $45,000, FICA is found by multiplying their gross pay for the pay period by 7.51%.

Example 1

Jennifer Slade earns $286 per week as a respiratory therapist. She has earned $11,219 so far this year. What is her FICA tax withheld?

1. Has she earned over $45,000? No, she has earned only $11,219.

2. Find FICA tax withheld. $286 × 7.51% = $21.48

Problems   Find each individual's FICA tax. The tax rate is 7.51% of the first $45,000 earned. .

		Earnings so far This Year	Current Gross Pay	FICA Tax/ Pay Period
1)	Dustin Rice	$38,215	$1,060	_____
2)	Steve Houston	15,146	349	_____
3)	Alica Meyer	14,978	1,215	_____
4)	Karen Bitler	45,810	2,614	_____
5)	Sheila Cooper	27,060	419	_____
6)	Candy Craig	11,015	216	_____
7)	Summer Hayes	41,263	2,117	_____
8)	Jack Tobias	17,651	396	_____
9)	Reagan Wise	1,214	1,214	_____
10)	Cora Tyler	10,054	391	_____

11) Cliff Steiner earns $614.89 per week as a dental hygienist. He has worked 39 weeks so far this year. What will be the FICA tax withheld during his 40th week of work?   _____

12) Janice Keith earns $7,298 per month as a pediatrician. What will be the FICA tax withheld during July?   _____

13) Kristin Russell earns $218.90 per week as a receptionist in Dr. Blair's office. She has earned $10,945 so far this year. What will be her FICA tax withheld this week?   _____

14) Sharon Dailey earns $9.57 per hour as a therapist at Oakland Day Care Center. She works 40 hours a week. She has earned $4593.60 so far this year. What will be her FICA tax withheld this week?   _____

15) Dale Bishop earns $16.95 per hour as a nursing home administrator. He works a 40-hour week. He has worked 50 weeks so far this year. What will be his FICA tax withheld during his 51st week of work?   _____

Voluntary Deductions

Other deductions that may be taken out of your paycheck include medical or life insurance, credit union deductions, union dues, charitable contributions, or other work-related memberships or expenses.

Insurance deductions vary depending on whether or not they are offered by your employer as a benefit. Often an employer will agree to pay part of your insurance cost as a benefit to you, while the remaining amount is deducted from your paycheck. The other voluntary deductions listed are usually taken from your check as a matter of choice and can vary in amount or frequency.

To find the amount deducted for group insurance, first find the percent of the cost that the employee must pay. You will usually be given the percent that the employer pays and you will use that information to determine what is left for the employee to pay. For example, if the employer pays 80% of the cost, that leaves 20% left for the employee to pay (100% – 80% = 20%).

Next, multiply the percent left for the employee to pay by the annual cost of the insurance. This will give you the amount that the employee pays annually for insurance.

Finally, divide that annual amount by the number of pay periods the employee works in a year. This will determine the amount deducted from his paycheck each pay period.

Example 1

Brenda Pope has medical and life insurance through her employer. The cost of the medical insurance is $1590 per year, and her employer pays 80%. The cost of the life insurance is $480 per year, and her employer pays 90%. How much is deducted from her biweekly paycheck for medical and life insurance?

Medical $1590 × 20% (100% – 80%) = $318.00
Life   480 × 10% (100% – 90%) =   48.00
                                      $366.00 annual amount

$366 divided by 26 (biweekly) = $14.08 deducted biweekly for insurance

Problems Determine how much is deducted from each paycheck for various insurances and other voluntary deductions.

1) Jay Brianco works at Technology Limited as a medical researcher. He has medical, life, dental, and vision insurance. He also has $15 per week deducted for AMA dues and $50 per week deducted to save in the credit union. His employer pays 70% of dental and vision and 90% of medical and life insurance. The annual costs are $1880 for medical, $1290 for life, $586 for dental, and $612 for vision.

   a. How much is deducted each week for medical insurance? _____

   b. How much is deducted each week for life insurance? _____

   c. How much is deducted each week for dental insurance? _____

   d. How much is deducted each week for vision insurance? _____

   e. What are the total voluntary deductions? _____

2) Chris Wheeler works at Princeville Vocational School as a school nurse. She earns $27,826 a year. She has medical and life insurance. She also has $10 deducted biweekly for the United Way. Medical insurance costs $1900 per year, and life insurance costs $1480 per year. Her employer pays 85% of all insurance costs.

   a. How much is deducted biweekly for medical insurance? _____

b. How much is deducted biweekly for life insurance? _____

c. What are the total voluntary deductions? _____

3) Cary Joseph is a veterinary's assistant at Hallman's Animal Hospital. He earns $20,050 a year and is paid monthly. His benefits include 90% coverage on medical and life insurance and 50% coverage on any other insurance plans. He has medical insurance that costs $2,098 a year, life that costs $1298 a year, vision for $950 a year, and a prescription drug card for $1105 per year. How much is deducted each month for

a. medical insurance? _____

b. life insurance? _____

c. vision insurance? _____

d. prescription insurance? _____

e. What are the total voluntary deductions? _____

4) Shawn Regent is an orderly at Howtonville Hospital. He has $50 a week deducted to deposit in his credit union, $5.50 a week deducted for union dues, $12.50 a week deducted for uniform fees, and $7.80 a week deducted for meals. He also has medical insurance for $16.58 a week and life for $3.21 a week.

a. What are the total voluntary deductions? _____

5) Mike Hawkins is a medical reporter for the KBSC news department. He earns $14,083 a year. He has medical insurance, life insurance, and travel insurance. He also has $10 deducted per week for deposit in his credit union. Medical insurance costs $1900 a year, life insurance costs $960 a year, and travel insurance costs $450 a year. KBSC pays 80% of all insurance.

a. How much is deducted each week for medical insurance? _____

b. How much is deducted each week for life insurance? _____

c. How much is deducted each week for travel insurance? _____

d. What are the total voluntary deductions? _____

6) Leigh Waxaw is an anesthesiologist. She has medical insurance through Parkview Medical Center's group plan. The cost of the insurance is $2926 a year, of which she pays 100%. How much does she pay each month for medical insurance? _____

7) Dana Sindell is a medical illustrator for Dove Publishers. She earns $28,717 a year. She has $25 per pay period deducted for a credit union loan, $6.95 per pay period deducted for membership fees, and $5 per pay period deducted for a flower fund. What is the total of her voluntary deductions each pay period? _____

8) Donna Watson is a receptionist at Dr. Douglas's office. She has medical insurance for $1815 a year and a prescription drug card for $1224 a year through her employer. The doctor pays 90% of all insurances.

a. How much is deducted each week for medical insurance? _____

b. How much is deducted each week for prescription insurance? _____

9) Lois Callahan works at Radel-Dowell Corp. as a medical/legal consultant. She has medical, life, dental, and vision insurance. She also has $15 per week deducted for AMA dues and $50 per week deducted to save in the credit union. Her employer pays 70% of the dental and vision costs and 90% of medical and life insurance costs. The annual costs are $2100 for medical, $965 for life, $495 for dental, and $705 for vision.

   a. How much is deducted each week for medical insurance?   _____

   b. How much is deducted each week for life insurance?   _____

   c. How much is deducted each week for dental insurance?   _____

   d. How much is deducted each week for vision insurance?   _____

   e. What is the total voluntary deductions?   _____

10) Randy Murphy is an orderly at Jefferson Hospital. He has $25/week deducted to deposit in his credit union, $4/week for union dues, $10.75/week for uniform fees, and $9.50/week for meals. He also has medical insurance for $21.15/week and life for $4.55/week.

   a. What are the total voluntary deductions?   _____

Net Pay

Net pay is the amount of pay earned for a given period of time after all deductions for taxes, insurances, and other voluntary deductions have been taken out. Net pay is the amount you receive on your paycheck.

To compute net pay, first find the individual's gross pay for the pay period. Next, find the sum of all deductions. Then:

GROSS PAY – DEDUCTIONS = NET PAY

Problems   Use Tables 15–1 and 15–2 on page 142 for FIT and find each individual's net pay.

1) Anna Deskart earns $6.08 per hour and double time for all overtime. Last week she worked 40 regular hours and 4 hours overtime. She is single, has 1 dependent, and claims two allowances. Her state tax rate is 4.05%. She has earned $10,496 so far this year.

   a. Find her gross pay for the week.   _____

   b. Find her FIT withheld for the week.   _____

   c. Find her state tax withheld per week.   _____

   d. Find her FICA tax withheld per week.   _____

   e. Find the total of all deductions.   _____

   f. Find her net pay.   _____

2) Phil Everly earns $19,208 a year as an EMT specialist. He gets paid weekly. He is married, has 3 dependents, and claims 0 allowances. His state tax rate is 2.5%. He has $15 per week deducted for deposit in his credit union and $2 per week deducted for membership dues.

   a. Find his gross pay for the week.   _____

   b. Find his FIT withheld for the week.   _____

   c. Find his state tax withheld per week.   _____

   d. Find his FICA tax withheld per week.   _____

  e. Find the total of all deductions.   _____

  f. Find his net pay.   _____

3) Cecil Mosgrove works as a physical therapist's aide and earns $5.70 per hour. He works 40 hours a week. He is married, has 2 children, and claims 4 allowances. His state tax rate is 4%, and he has worked 27 weeks so far this year. He has $2.80 deducted per week for union dues.

  a. Find his gross pay for the week.   _____

  b. Find his FIT withheld for the week.   _____

  c. How much has he earned so far this year?   _____

  d. Find his state tax withheld per week.   _____

  e. Find his FICA tax withheld per week.   _____

  f. Find the total of all deductions.   _____

  g. Find his net pay.   _____

4) Jana Wyler earns $19,826 a year as an LPN instructor at Highland Technical School. She is married with no dependents. She claims no allowances. Her state does not assess a tax.

  a. Find her gross pay for the week.   _____

  b. Find her FIT withheld for the week.   _____

  c. Find her state tax withheld per week.   _____

  d. Find her FICA tax withheld per week.   _____

  e. Find the total of all deductions.   _____

  f. Find her net pay.   _____

5) Audrey Banks works as a housekeeper at Ashville Hills Elderly Care Center. She earns $5.25 per hour and works 40 hours a week. She is single with no dependents and claims 0 allowances. Her state tax rate is 5.25%.

  a. Find her gross pay for the week.   _____

  b. Find her FIT withheld for the week.   _____

  c. Find her state tax withheld per week.   _____

  d. Find her FICA tax withheld per week.   _____

  e. Find the total of all deductions.   _____

  f. Find her net pay.   _____

Earning Statement

An earning statement is a record of an employee's earnings, deductions, and net pay. Often it also contains a record of the individual's earnings and deductions so far during the year (earnings to date).

The earning statement is often attached to the individual's paycheck and is sometimes called a check stub. The earning statement and the paycheck itself can be done by hand, by computer, or on a write-it-once system like Control-O-Fax (see figure 15–3).

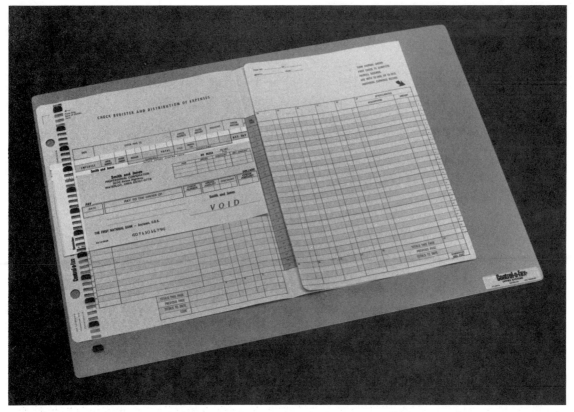

Figure 15–3 **Payroll check system** *(Courtesy of Control-O-Fax Office Systems, Waterloo, IA)*

Problems Complete the following earning statements using the problems solved in the previous section of this workbook (net pay problems 1–5).

1) Employee: Anna Deskart
   S.S.# 211–41–8112

Name		S.S.#		Pay Rate	
FIT	FICA	State	Other	Total Ded.	
Reg. Pay	Overtime Pay	Gross Pay		Net Pay	

2) Employee: Phil Everly
   S.S.# 196–06–8621

Name		S.S.#		Pay Rate	
FIT	FICA	State	Other	Total Ded.	
Reg. Pay	Overtime Pay	Gross Pay		Net Pay	

3) Employee: Cecil Mosgrove
S.S.# 313–13–3113

Name _____	Reg. Pay _____
S.S.# _____	Overtime Pay _____
Pay Rate _____	
THIS PAY PERIOD	**YEAR TO DATE**
Gross pay _____	Gross pay _____
FIT _____	FIT _____
State _____	State _____
FICA _____	FICA _____
Other _____	Other _____
Net Pay _____	Net Pay _____

4) Employee: Jana Wyler
S.S.# 276–01–0455

Name _____	Reg. Pay _____
S.S.# _____	Overtime Pay _____
Pay Rate _____	
THIS PAY PERIOD	**YEAR TO DATE**
Gross pay _____	Gross pay _____
FIT _____	FIT _____
State _____	State _____
FICA _____	FICA _____
Other _____	Other _____
Net Pay _____	Net Pay _____

5) Employee: Audrey Banks
S.S.# 416–00–5523

Name		S.S.#		Pay Rate	
FIT	FICA	State	Other		Total Ded.
Reg. Pay	Overtime Pay	Gross Pay		Net Pay	

Paychecks

To write a paycheck, most offices or companies use an official payroll check rather than one from their normal checking account. It is necessary to have an authorized signature on all checks written. Sometimes, the one completing the payroll is given the authority to sign the checks, and other times, a doctor, office manager, or other supervisor may sign the checks. In these exercises, you have the authority to sign the checks.

To complete a check, fill in the date it is being written, the name of the employee, the amount of net pay in numbers and also in words (as explained in chapter 1). Then, sign the check. If an earning statement is attached, complete it as well.

Problems Complete a paycheck for each employee in the previous section.

1)

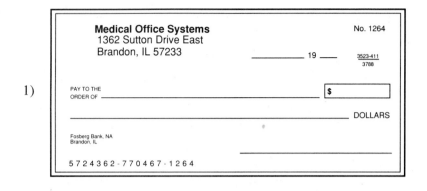

Medical Office Systems
1362 Sutton Drive East
Brandon, IL 57233

No. 1264

_____ 19 ___   3523-411
                 3788

PAY TO THE
ORDER OF _____ $ _____

_____ DOLLARS

Fosberg Bank, NA
Brandon, IL

5724362-770467-1264

2)

Medical Office Systems
1362 Sutton Drive East
Brandon, IL 57233

No. 1265

_____ 19 ___   3523-411
                 3788

PAY TO THE
ORDER OF _____ $ _____

_____ DOLLARS

Fosberg Bank, NA
Brandon, IL

5724362-770467-1265

3)

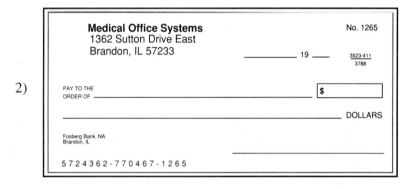

Medical Office Systems
1362 Sutton Drive East
Brandon, IL 57233

No. 1266

_____ 19 ___   3523-411
                 3788

PAY TO THE
ORDER OF _____ $ _____

_____ DOLLARS

Fosberg Bank, NA
Brandon, IL

5724362-770467-1266

4)

Medical Office Systems
1362 Sutton Drive East
Brandon, IL 57233

No. 1267

_____ 19 ___   3523-411
                 3788

PAY TO THE
ORDER OF _____ $ _____

_____ DOLLARS

Fosberg Bank, NA
Brandon, IL

5724362-770467-1267

5)

```
┌─────────────────────────────────────────────────────────────┐
│   Medical Office Systems                      No. 1268        │
│   1362 Sutton Drive East                                      │
│   Brandon, IL 57233            _____ 19 ___    3523-411     │
│                                                     3788       │
│                                                               │
│   PAY TO THE                                  ┌─────────────┐ │
│   ORDER OF  _____  │ $           │ │
│                                               └─────────────┘ │
│   _____  DOLLARS        │
│                                                               │
│   Fosberg Bank, NA                                            │
│   Brandon, IL                  _____  │
│                                                               │
│   5724362-770467-1268                                         │
└─────────────────────────────────────────────────────────────┘
```

In problems 6–10, complete each week's earning statement and paycheck for George Prang (S.S.# 017–17–2416). He is married, has 3 children, and claims all allowances he is entitled to claim. His state tax rate is $2\frac{3}{4}\%$ with no exemptions allowed. He has earned $18,106 so far this year. He earns $7.80 an hour, with time and a half for overtime.

6) George worked 40 regular hours and 6 hours overtime. He has $22.01 deducted this week for health insurance and $25.00 deducted for union dues.

Name		S.S.#		Pay Rate	
FIT	FICA	State	Other		Total Ded.
Reg. Pay	Overtime Pay	Gross Pay		Net Pay	

```
┌─────────────────────────────────────────────────────────────┐
│   Medical Office Systems                      No. 1264        │
│   1362 Sutton Drive East                                      │
│   Brandon, IL 57233            _____ 19 ___    3523-411     │
│                                                     3788       │
│                                                               │
│   PAY TO THE                                  ┌─────────────┐ │
│   ORDER OF  _____  │ $           │ │
│                                               └─────────────┘ │
│   _____  DOLLARS        │
│                                                               │
│   Fosberg Bank, NA                                            │
│   Brandon, IL                  _____  │
│                                                               │
│   5724362-770467-1264                                         │
└─────────────────────────────────────────────────────────────┘
```

7) George worked 40 regular hours and $3\frac{1}{2}$ hours overtime.

Name		S.S.#		Pay Rate	
FIT	FICA	State	Other		Total Ded.
Reg. Pay	Overtime Pay	Gross Pay		Net Pay	

```
┌─────────────────────────────────────────────────────────────┐
│   Medical Office Systems                      No. 1265        │
│   1362 Sutton Drive East                                      │
│   Brandon, IL 57233            _____ 19 ___    3523-411     │
│                                                     3788       │
│                                                               │
│   PAY TO THE                                  ┌─────────────┐ │
│   ORDER OF  _____  │ $           │ │
│                                               └─────────────┘ │
│   _____  DOLLARS        │
│                                                               │
│   Fosberg Bank, NA                                            │
│   Brandon, IL                  _____  │
│                                                               │
│   5724362-770467-1265                                         │
└─────────────────────────────────────────────────────────────┘
```

8) George worked 40 regular hours. He had $22.01 deducted this week for health insurance.

Name		S.S.#		Pay Rate	
FIT	FICA	State	Other	Total Ded.	
Reg. Pay	Overtime Pay	Gross Pay		Net Pay	

Medical Office Systems
1362 Sutton Drive East
Brandon, IL 57233

No. 1266

_____ 19 _____   3523-411
                                 3788

PAY TO THE
ORDER OF _____ $ _____

_____ DOLLARS

Fosberg Bank, NA
Brandon, IL

5724362·770467·1266

9) George worked 40 regular hours and $5\frac{1}{4}$ hours overtime.

Name		S.S.#		Pay Rate	
FIT	FICA	State	Other	Total Ded.	
Reg. Pay	Overtime Pay	Gross Pay		Net Pay	

Medical Office Systems
1362 Sutton Drive East
Brandon, IL 57233

No. 1267

_____ 19 _____   3523-411
                                 3788

PAY TO THE
ORDER OF _____ $ _____

_____ DOLLARS

Fosberg Bank, NA
Brandon, IL

5724362·770467·1267

10) George worked 40 regular hours and 1 hour overtime. He had $22.01 deducted for health insurance and $25.00 for union dues.

Name		S.S.#		Pay Rate	
FIT	FICA	State	Other	Total Ded.	
Reg. Pay	Overtime Pay	Gross Pay		Net Pay	

Medical Office Systems
1362 Sutton Drive East
Brandon, IL 57233

No. 1268

_____ 19 _____   3523-411
                                 3788

PAY TO THE
ORDER OF _____ $ _____

_____ DOLLARS

Fosberg Bank, NA
Brandon, IL

5724362·770467·1268

Chapter 16

Office Skills

Numerical Filing

There are two basic types of numerical filing: terminal number filing and regular filing. **Terminal number** filing is a system where each file is numbered with a series of digits followed by a set of trailing digits. An example of a terminal number is 1637–55. In this case, the file number is 1637 and the terminal (ending) digits are 55. **Regular** filing is a system where each file is numbered with an uninterrupted series of digits. An example of a regular file number is 1637. There is just the one series of numbers.

There are a few basic rules in numerical filing.

1. Treat all numbers being filed as though they contained the same number of digits. For example, if filing 15 and 1747, treat 15 as a four-digit number, 0015. Then proceed with filing.

2. Once you consider all numbers to be the same length, compare the numbers left to right (as discussed in chapter 1). File the smaller numbers first, then the larger ones.

3. In terminal number filing, sort the files by terminal (ending) digits first. Then, put each set containing similar terminal digits in numerical order according to the rules above. (See figure 16–1.)

Terminal numbers are used with files for various reasons. Some offices use terminal numbers on their files to distinguish which doctor the patient belongs to in the practice. Terminal numbers can also be used to determine the age of an account when used with your accounting files. In a pediatrics office, terminal numbers may be used on files to show the year of birth of the patient (e.g., 1637–90 indicates a patient born in 1990, etc.). Each office may have a different use for terminal numbers but, regardless of the use, the system is very helpful in sorting different types of files within your filing system.

To file regular files, simply put them in numerical order according to rules 1 and 2 listed above. To file terminal number files, sort the files by terminal number first (as explained in rule 3). Then, put each group in numerical order.

Example 1

Put the following files in order.

1536	17	1758	14	175	189
1637	177	1759	141	198	1677

1. First, treat all numbers being filed as though they contain the same number of digits. The longest numbers have four digits in them, so treat all numbers as four-digit numbers.

1536	**0017**	1758	**0014**	0175	0189
1637	**0177**	1759	**0141**	0198	1677

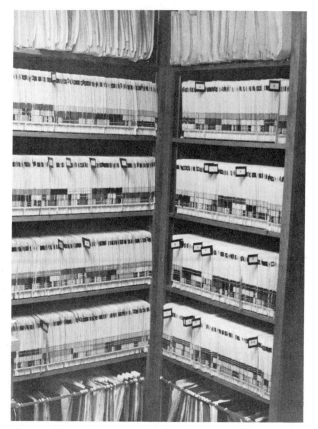

Figure 16–1 Color-coded file room. Reprinted with permission from Simmers, *Diversified Health Occupations*, 2E, copyright 1983, Delmar Publishers, Albany, NY.

2. Next, compare the numbers left to right and put them in order from smallest to largest.

0014	Numbers beginning in 00 come first: 0014 before 0017.
0017	
0141	Numbers beginning in 0 come next, in order.
0175	
0177	
0189	
0198	
1536	Numbers beginning in 1 come next: 15—, 16—, then 17—.
1637	
1677	
1758	
1759	

Example 2

Put the following files in numerical order.

163–6	183–3	155–3	123–3	156–3	108–3
167–3	155–6	151–6	109–6	160–6	111–3
163–3	162–6	177–6	148–3	16–6	171–6
174–6	145–3				

1. First, sort the different terminal number groups.

__-3__	__-6__	Notice that there is no special order to these numbers; you are
167–3	163–6	just separating the 3s from the 6s at this point.
163–3	174–6	
183–3	155–6	
145–3	162–6	
155–3	151–6	
123–3	177–6	
148–3	109–6	
156–3	160–6	
108–3	016–6	
111–3	171–6	

2. Put the 3s in numerical order. All begin in 1, so look at the second digit.

108–3	
111–3	
123–3	
145–3	145 before 148
148–3	
155–3	155 before 156
156–3	
163–3	163 before 167
167–3	
183–3	

3. Next, put the 6s in numerical order. One begins in zero, the rest begin with 1.

016–6	
109–6	
151–6	151 before 155
155–6	
160–6	160, then 162, then 163
162–6	
163–6	
171–6	171, then 174, then 177
174–6	
177–6	

4. The final filing order (both groups combined with the 3s before the 6s) is written down and to the right:

108–3	148–3	167–3	151–6	163–6
111–3	155–3	183–3	155–6	171–6
123–3	156–3	016–6	160–6	174–6
145–3	163–3	109–6	162–6	177–6

Sometimes different terminal number groups are filed in the same set of filing cabinets, as would be the case in the example of the pediatrics office. All patients files are kept together in the same room, but they are sorted by the year of birth. In other cases, the different groups may be filed in entirely different places. For example, one doctor's files, coded –55, may be kept in a separate room from another doctor's files, coded –50. It is important to follow the filing process that is used in each particular office or hospital you may work in.

Problems Put the following groups of files in numerical order.

1)	1776	167	177	1889	1361
	1637	16371	19003	1899	1248
	17482	1748	17417	1714	1137
	174	1763	1741	18292	13781
	189	1822	1737	176	1988
	11718	1828	18229	1789	12745
	198	16370			

2)	382	247	374	347	293
	398	294	546	482	193
	396	399	732	721	483
	492	583	932	117	822
	572	488	568	653	542
	278	472	113	441	611
	713	134	17	346	764

3)	1713741	2345654	1374317	1267537	2345676
	1347431	2312341	1473134	1347413	3457834
	1345435	1236431	3456765	1347874	1234564
	1347123	1373483	2347245	1347871	2347423
	1378976	1347434	2345511	1384134	

4)	153–65	135–67	163–66	172–66	263–65
	123–66	126–67	123–65	127–67	237–66
	173–66	237–65	237–66	231–65	284–67
	272–67	832–65	234–67	278–66	181–66
	173–66	236–65	211–66	171–65	288–67

5)	16134–6	12742–6	17437–6	18324–5	12367–5
	12362–5	12728–6	1236–5	12473–5	17347–6
	12362–6	17437–6	1263–6	12754–6	12347–5
	12742–5	12363–5	13873–5	17246–5	18534–6
	16326–6				

6)	12	1853	1236	17443	1274
	174	183	0123	1623	1952
	12378	163	06236	723	952
	1236	16734	12333	2347	9952
	127				

Time Management (appointment scheduling)

Time management is very important in the medical office. **Time management** is the scheduling of apointments, breaks, and various other duties for a physician, dentist, veterinarian, or other individual.

First, let's look at an appointment book. Appointment books may have more than one column for each day—one for each different physician—or they may have only one in a single-doctor practice. The time may be divided in 15-minute intervals or in intervals of a half an hour. The specific appointment book that we will use in our example has several outstanding features:

different colored pages for each day of the week

tabs on each page that show the 15-minute intervals, which can then be shaded in to show for a week at a time which time slots are filled

a six-day plan for offices that might open on Saturdays

hole-punched for use in a three-ring binder; this helps offices who need to schedule appointments six months or a year in advance by always keeping the book filled with new pages for a year in advance (old pages can be taken out and kept on file elsewhere)

This appointment book is designed by Control-O-Fax (see figure 16–2)

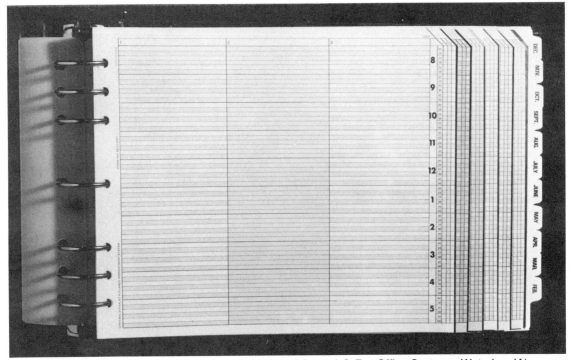

Figure 16–2   Appointment book *(Courtesy of Control-O-Fax Office Systems, Waterloo, IA)*

When you record an apointment, you should write down the following information on the correct appointment line:

patient's name   person responsible for paying the bill
telephone number   reason for office visit

You must also symbolize how long the appointment will last by drawing a line down to the appropriate time and by crossing out the time slots on the far right tab.

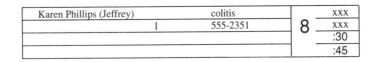

Figure 16–3

This is an appointment for Karen Phillips, who has colitis. Jeffrey (Phillips, her husband) is responsible for paying the bill. This appointment should take a half hour, so it takes up two 15-minute intervals in the appointment book, thus, cross off 8:00 and 8:15. That leaves 8:30 as the next available time for scheduling an appointment (see figure 16–3).

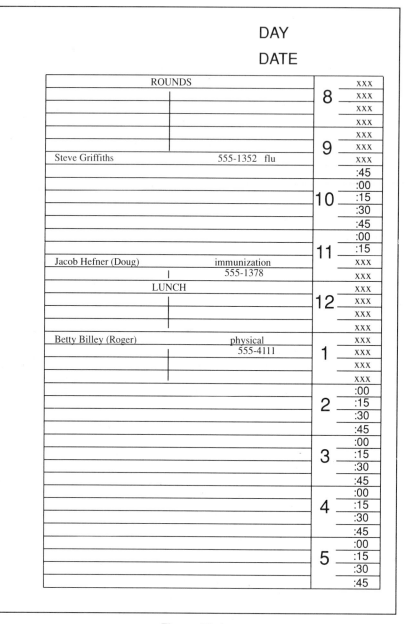

Figure 16–4

Notice that lunch is scheduled from 12:00 until 1:00 and that the doctor makes rounds at the hospital from 8:00 until 9:30. This day also has three patients scheduled so far—one at 9:30 (which lasts 15 minutes), one at 11:30 (which lasts a half hour), and one at 1:00 (which lasts an hour). You can glance down the right side and see which times are available and which are not.

Remember, each line represents 15 minutes. If an appointment takes 15 minutes, give one line only. If it takes a half hour, give two lines. Forty-five minutes equals three lines, and an hour equals four lines. Be sure to draw a line when necessary and always to cross off the time slots on the right side.

Problems   Complete the following two-day appointment book. Schedule all appointments as early as possible, staying within the time the patient requests. (If they ask for after lunch, give the first time available after lunch.) Also, once a time limit is established, it remains the same time for the rest of the problem (example: if a physical takes an hour for one person, it will take that long for everyone—unless otherwise indicated).

1) Label the first day Wednesday, February 27, and the second day Thursday, February 28.

2) The doctor has rounds at the hospital every day until 9:00. (**Until** means that he will be back in the office **at** 9:00.)

3) Lunch is taken from 12:30 to 1:30 every day. On Thursday, the doctor has a luncheon meeting that will last until 2:30.

4) The doctor has an appointment with his optometrist on Wednesday at 10:00. This should last 30 minutes.

5) Mrs. Karen Nobley calls the doctor to get an appointment for her son, Frank. He needs a physical. This should last one hour. She prefers Wednesday at 3:00 pm. She is responsible for this bill (555–1234).

6) Mr. Don Edwards needs an appointment to have his blood pressure checked. This requires 15 minutes and he would like to come in *right before lunch* on Thursday (555–4939).

7) Miss Jackie Stein needs to come in for an allergy shot. This requires 15 minutes, and she would like early Wednesday morning (555–0913).

8) Mrs. Edith Harrington needs to bring her twins in for their yearly checkup. This will require 45 minutes. She prefers 11:00 Wednesday morning (555–0356).

9) Mrs. Celia Davis needs a physical. She prefers early Thursday morning (555–4873).

10) Mr. Walter Sackston needs his blood pressure checked. He prefers right after lunch on Wednesday (555–1362).

11) Mrs. Corinne McDaniel wants to bring her daughter in because she has a cold. This will require 15 minutes. She prefers 3:45 either day. Leslie is her daughter's name (555–8316).

12) Mrs. Tracy Hart needs a Pap test. This will require 15 minutes. She prefers right after lunch on Thursday (555–3761).

13) Mrs. Doris Haybron needs to have tests done for allergies. This will require a half hour. She prefers noon on Wednesday (555–3168).

14) Mrs. Phyllis Hudson needs a Pap test. She prefers early Thursday morning (no phone).

15) Mr. Wayne McGregor wants to get a physical. He wants the first available opening; it doesn't matter which day (555–7777).

16) Miss Lena Hoehen needs to come in for a skin rash. She prefers Wednesday morning after 11:00. This requires 15 minutes (555–4787).

17) Mr. Justin Landsbury wants to have his burned arm checked out. He has seen the doctor about this earlier and needs only a brief checkup. This should take 15 minutes. The best time for him is 4:30 Thursday (555–7070).

18) Mrs. Marie Hayes needs a physical. She can come only on Thursday afternoon (555–1378).

19) Mrs. Lorraine Good would like to bring her twins in for bronchitis. This will require 30 minutes. She can come early Wednesday afternoon (555–0113).

20) Mr. Sidney Shillings needs a physical. He can come anytime (555–3636) (remember: earliest openings first).

21) Mrs. Diane Doyle needs to have her blood pressure checked. She prefers 9:30 either day (555–0411).

22) Ms. Harriett Jolson needs to come in for an allergy shot. Anytime after 9:45 Wednesday morning is all right (555–3101).

23) Mr. Isaac Klein needs to meet with the doctor Wednesday afternoon for a newspaper interview. This should require about 45 minutes, and the latest time in the day would be best (*High Plains Dealer*, 555–8765).

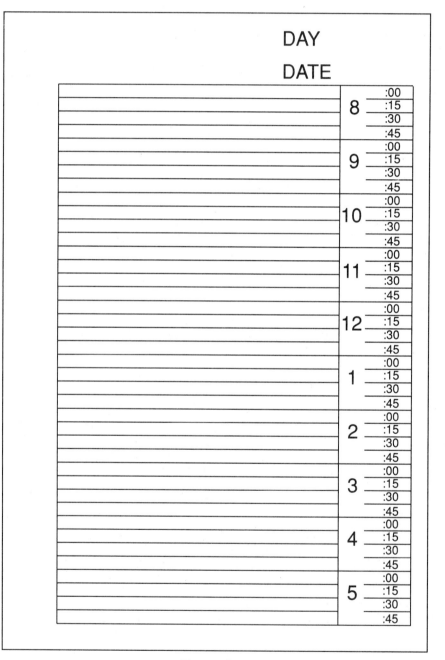

Figure 16–5

DAY

DATE

	8	:00
		:15
		:30
		:45
	9	:00
		:15
		:30
		:45
	10	:00
		:15
		:30
		:45
	11	:00
		:15
		:30
		:45
	12	:00
		:15
		:30
		:45
	1	:00
		:15
		:30
		:45
	2	:00
		:15
		:30
		:45
	3	:00
		:15
		:30
		:45
	4	:00
		:15
		:30
		:45
	5	:00
		:15
		:30
		:45

Writing Receipts

There are three basic types of receipts written in a medical office. The first comes from a handwritten bookkeeping system that some offices use. This is a form that varies in size from $8\frac{1}{2}$" by 5" to as large as $8\frac{1}{2}$" by 11". It contains spaces for the following information:

date
name of patient
previous balance
new charges
current balance

diagnosis
services performed
next appointment
other information needed by insurance

This type of receipt is complete enough that it can be submitted to the insurance company for payment of the bill (along with the regular insurance form that the patient is required to complete for his insurance company) (see figure 16–6).

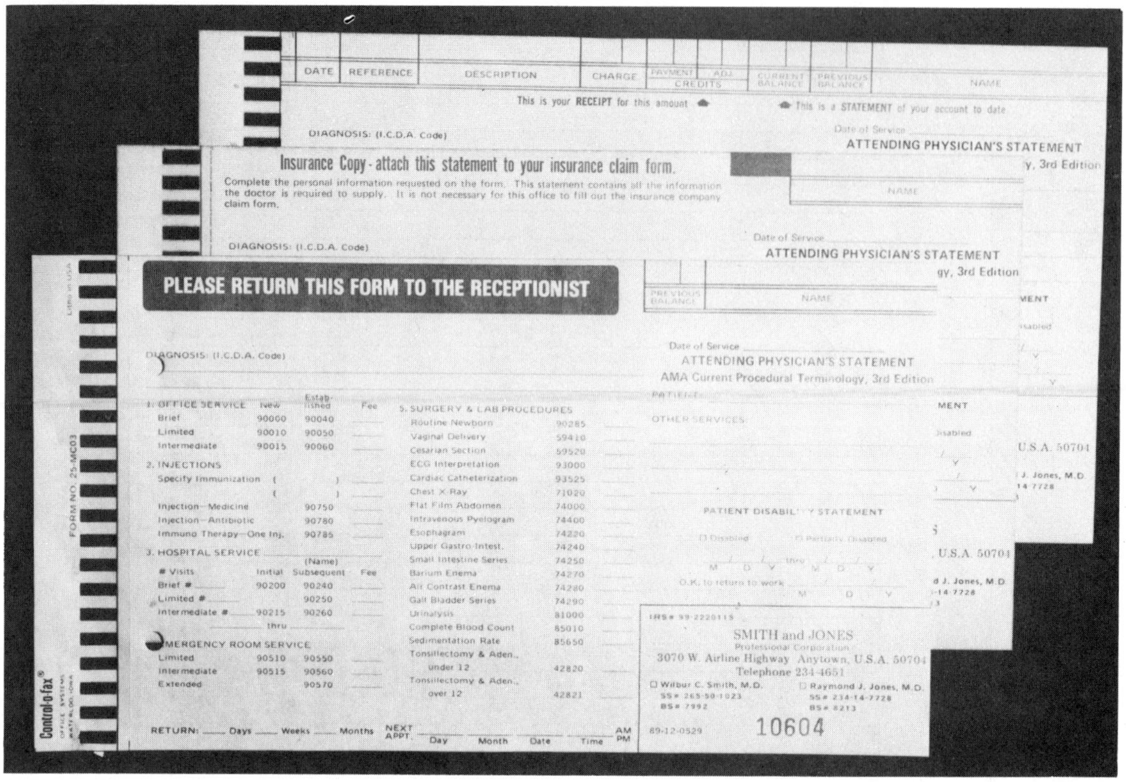

Figure 16–6 Super bill *(Courtesy of Control-O-Fax Office Systems, Waterloo, IA)*

The next type is a computer-generated receipt. This type is similar in content to the handwritten type, but all of the information is typed into a computer and the receipt is automatically produced. This receipt is just a small part of a large data (information) base that contains patient files, billing information, and scheduling, all of which are on a computer. What you type in for the patient's receipt goes automatically into his medical file and into his billing file (see figure 16–7).

Adrian Animal Hospital, 1657 Highway East, Lima, OH 45801

Kathi Dunlap
2145 N. Dixie
Lima, OH 45801

Account No: 3380

Patient: Lindsey — Species: Canine — Sex: FE
Breed: Small Mixed Breed — Age: 9w

	Qty.	U.Price	Total
Fecal Examination-Parasites	1	7.10	7.10
DHCP-Parvo 1st of 3	1	19.50	19.50
Albon Liquid 1 oz bottle	1	4.35	4.35
Heartgard 68 mcg tablets	1	12.10	12.10

Previous Balance	0.00	Total Invoice	43.05
Invoice 15759	43.05	Sales Tax	0.00
Sub Total	43.05	Net Invoice	43.05
Payment	43.05		
Balance Due	0.00		

Figure 16–7

The third type is a handwritten receipt that records only the money transacted. It does not give information about the diagnosis or services and cannot be used for insurance purposes (see figure 16–8).

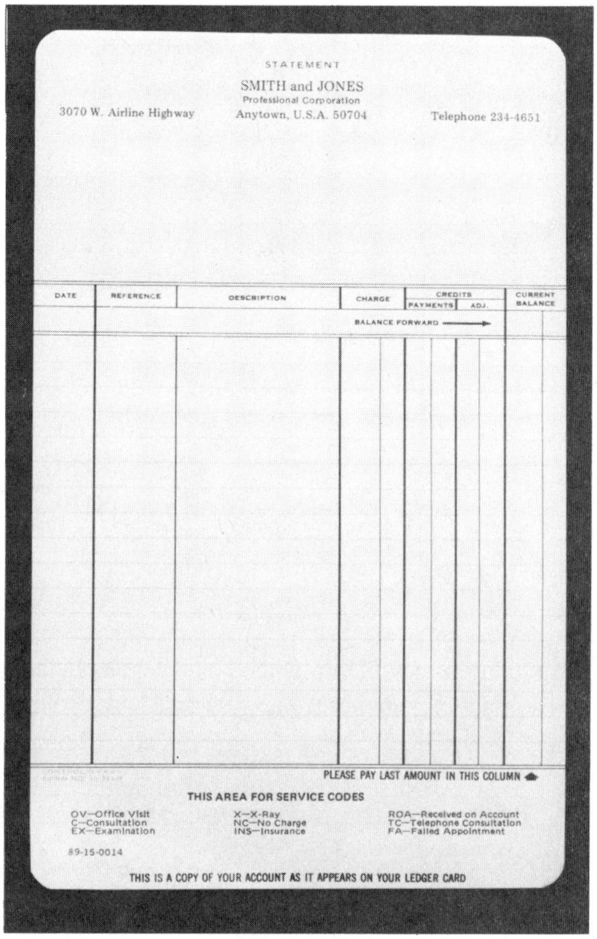
STATEMENT

SMITH and JONES
Professional Corporation

3070 W. Airline Highway | Anytown, U.S.A. 50704 | Telephone 234-4651

DATE	REFERENCE	DESCRIPTION	CHARGE	CREDITS		CURRENT BALANCE
				PAYMENTS	ADJ.	
			BALANCE FORWARD →			

PLEASE PAY LAST AMOUNT IN THIS COLUMN

THIS AREA FOR SERVICE CODES

OV—Office Visit
C—Consultation
EX—Examination
X—X-Ray
NC—No Charge
INS—Insurance
ROA—Received on Account
TC—Telephone Consultation
FA—Failed Appointment

89-15-0014

THIS IS A COPY OF YOUR ACCOUNT AS IT APPEARS ON YOUR LEDGER CARD

Figure 16–8 *(Courtesy of Control-O-Fax Office Systems, Waterloo, IA)*

When writing receipts, you must be able to figure out what the patient owes. This is called the **current balance**. To find his current balance, use the following formula:

CURRENT BALANCE = PREVIOUS BALANCE + CHARGE FOR TODAY'S SERVICES – PAYMENTS OR CREDITS

The patient's ledger card (his financial record with your office) shows the previous balance. The charge for today's services can be found by consulting the physician or by using a chart of customary charges used by your office. He should be told that they owe the total of the previous balance and today's charges, and they should be encouraged to pay the entire amount (see chapter 13 for the collection of money). Then, you subtract the amount that they pay you from this total and the part that remains (if anything) is their **current balance**.

Problems Complete the following information on the receipts. Be sure to sign your name on the receipts where this is called for.

1) Jack Elliott came into your office for a complete physical. The charge for this is $25, and he has a previous balance of $42. Today's date is 8/17. He is paying you $40 today.

Date	Patient	Charge	Payment	Cur. Bal.	Prev. Bal.

Description of Services _____

2) Nola Leven sent a payment of $415.25 to the office today for her recent stay in the hospital. Her previous balance is $915.25. Today's date is 7/11/90. She would like you to send her a receipt.

Date	Patient	Charge	Payment	Cur. Bal.	Prev. Bal.

Description of Services _____

3) Karen Hines came in today for an allergy shot. The charge for that is $16.50. Her previous balance is $0. She made a payment of $16.50. Today's date is 4/15/90.

Date	Patient	Charge	Payment	Cur. Bal.	Prev. Bal.

Description of Services _____

4) Greg Bercoff came in today for swollen glands. His previous balance is $19.00, and today's charges are $19.00. He made no payment. Today's date is 4/17/90.

Date	Patient	Charge	Payment	Cur. Bal.	Prev. Bal.

Description of Services _____

5) Sarah Mullenston came in today for pains in her side and was admitted to the hospital for an appendectomy. Her total bill will not be computed until she is released, but she did pay her previous balance of $27.00 when she came in. Write her a receipt. Today's date is 5/18/90.

Date	Patient	Charge	Payment	Cur. Bal.	Prev. Bal.

Description of Services _____

6) Hugh Golliver came in today and paid $50 on his doctor's bill. His previous balance was $163.55. Today's date is 11/8/90.

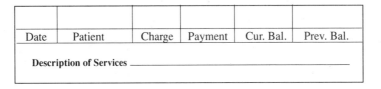

Date	Patient	Charge	Payment	Cur. Bal.	Prev. Bal.

Description of Services _____

Received of _____ $_____

_____ DOLLARS

_____

7) Randy Cramer came in today for some allergy tests. The charges are $45.50. His previous balance is $15.00. He is paying $10.00 today (10/6/90).

Date	Patient	Charge	Payment	Cur. Bal.	Prev. Bal.

Description of Services _____

Received of _____ $_____

_____ DOLLARS

_____

8) Donna Myers came in today because she has the flu. She owes $22.00 for today's services and has a previous balance of $15.00. She wants to pay the entire amount today (5/5/90).

Date	Patient	Charge	Payment	Cur. Bal.	Prev. Bal.

Description of Services _____

Received of _____ $_____

_____ DOLLARS

_____

9) Sean Sanders came in today for an athletic physical. The charge for that is $25.00 and he has no previous balance. He pays $25.00 today (9/9/90).

Date	Patient	Charge	Payment	Cur. Bal.	Prev. Bal.

Description of Services _____

10) Marie Hill came in today for a mark on her face. The doctor ordered a biopsy and sent her to the outpatient clinic for the lab work. She had a previous balance of $14.00, and her charge for today's visit is $22.00. She is paying nothing today.

Date	Patient	Charge	Payment	Cur. Bal.	Prev. Bal.

Description of Services _____

Chapter 17

Personal Finance

Working in the health profession can provide many benefits. A professional work environment, good wages and benefits, and the ability to provide care for the public are some of the benefits of the career you have chosen. The financial benefits should not be overlooked, since they will be important as you provide for yourself and your family.

Managing your money is not difficult, but it does take some self-discipline and some knowledge. There are many opportunities in life to spend money wisely, and unwisely. It is up to you to spend your money wisely, to save enough to provide for emergencies or future goals, and to quickly pay those to whom you owe money.

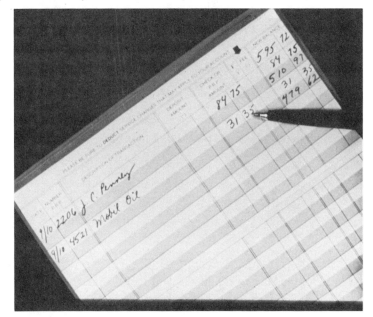

Figure 17-1

This chapter will look at checking accounts and savings accounts, which are tools to help you spend and save your money. Then we will look at credit (obtaining credit, using it, and paying the bills that are incurred). Next, we will examine various monthly expenditures that arise when you support yourself (utility bills and groceries), purchasing a car and making payments on it, and renting an apartment. Lastly, we will put all of the expenses together in a monthly budget.

Personal Checking

Having a checking account can be very useful when you have a steady income and monthly expenses. Writing a check is the same as paying cash for an item, but it is safer to send through the mail and it eliminates the need to carry large amounts of cash with you.

To open a checking account, you must first select a bank. Each bank offers different types of checking accounts—some that pay interest, some that have service charges (fees that you pay for special services the bank provides), some that have an overdraft protection (a certain amount of credit attached to your checking account that prevents a check from bouncing if you do not have enough money in your account), some that require a minimum balance (of $200, $500, or more), and some that offer extra services. Once you have selected a bank in which to open your account, you must go to that bank and fill out the appropriate application and forms for opening a checking account. You must deposit a certain amount of money (at least the minimum specified by that bank). Then, most banks will give you a set of beginner's checks to use until your checks are printed and sent to you.

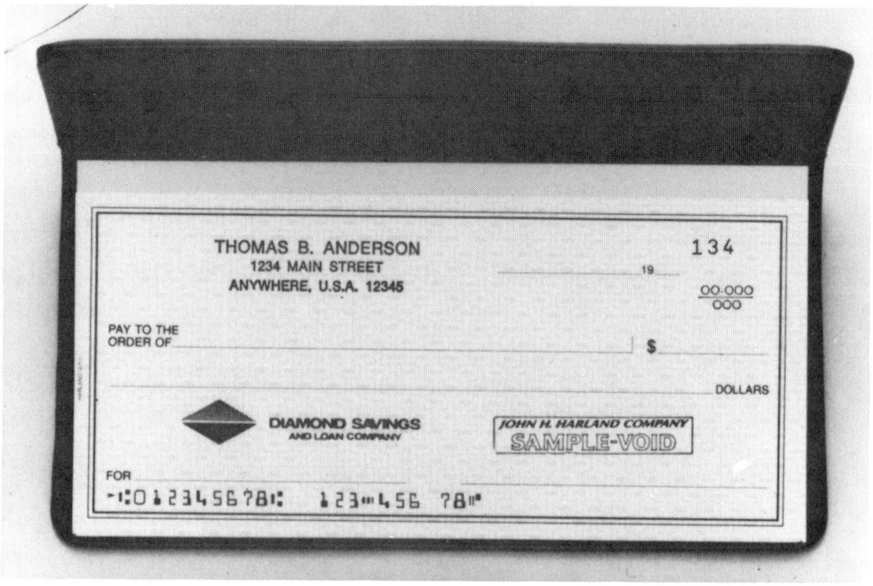

Figure 17–2

Once you have opened your checking account, you are authorized to begin writing checks from the amount you have deposited. To write a check, you must fill in the following:

the current date

the name of the person or organization to whom you are writing the check (called the payee)

the amount of the check in numbers

the amount of the check in words

your signature

some checks have a memo line where you can make a brief notation to yourself about the purpose of the check

When the check is written, it is wise to immediately record that check in the **check register**. This is a record of how much money you currently have in your checking account, according to what you have deposited and what you have spent. To record a check in the check register, you must fill in all of the information about the check and subtract the amount of the check from the balance to determine your new checking account balance.

Many banks now have automated teller machines (ATMs) that allow you to draw money out of your account during or after business hours. To use an ATM, you need to have an ATM card and a personal identification number (PIN). You insert the ATM card in the ATM and it asks you to enter your PIN. This number should be memorized and guarded carefully, since it is what protects your card from being misused by someone else. When you have correctly entered your identification number, you can withdraw money. A check does not need to be written, but you still must record the transaction in your check register and subtract the amount of the withdrawal from the balance to determine your new checking account balance.

You will also make deposits to your account. To make a deposit, you can go to your bank during business hours, use an ATM, or use a night deposit slot (available at most banks). You must fill out a deposit slip, indicating what you plan to deposit, and you must sign this deposit slip if you wish to receive any cash back from what you are submitting for deposit. For example, when you go to the bank with your payroll check, you may wish to deposit only $100 and you would like the rest back in cash. In this case, your deposit slip must be signed before they will give you the cash back. You should only use the night deposit when you wish to deposit the entire amount, since it will not be processed until the next business day. The deposit should also be recorded into your check register and the amount added to the balance to determine your new checking account balance.

Writing a personal check is the same as writing a check for your office (discussed in chapter 15 along with payroll checks). To write a deposit slip (see figure 17–3.), fill in the following:

- the date
- your account number
- the amount brought to the bank (usually broken down into coins, currency, and checks)
- the amount you wish to receive in cash
- the amount of the deposit

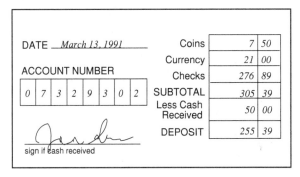

Figure 17–3

To fill in the check register for a check or a deposit, you must fill in all information requested. A check register usually looks something like this:

Number	Date	Description of Transaction	Payment	x	Deposit	Balance

Figure 17–4

The number means the number of the check (found in upper-right-hand corner); the date is the date the check was written; the description of transaction is who the check was written to as well as a description of its purpose, if desired; the payment column is for checks or ATM withdrawals or bank fees that come out of your account; the deposit column is for deposits or interest earned on the account that goes into your account; and the balance column records the amount currently in your account.

The bank will send you a statement of your account every month This statement records the deposits, the checks paid out from your account, and the interest earned (or fee charged). They also send you your **canceled checks**. These are checks that have been written from your account, cashed by the individual the check was written to, and returned to your bank for payment. They are marked "canceled" and are proof of the payment you made. They cannot be used again but should be kept for your records along with the statement from the bank.

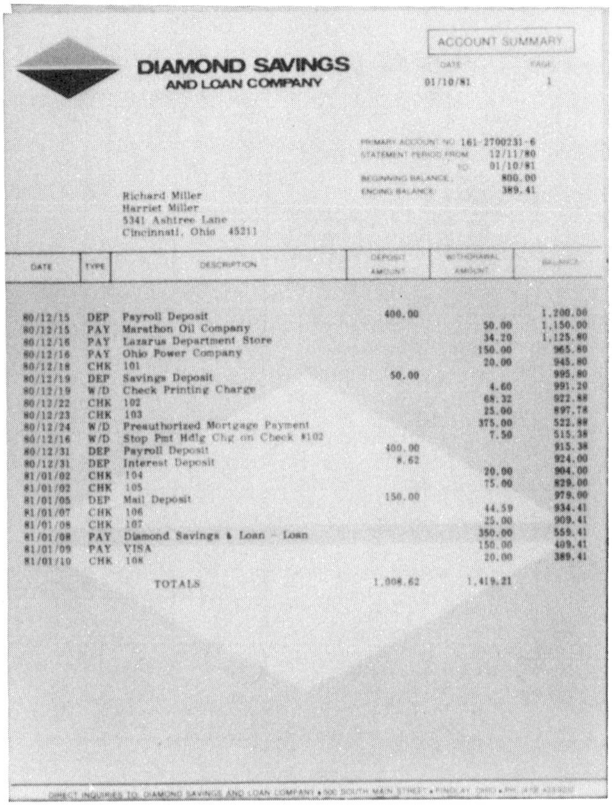

DIAMOND SAVINGS AND LOAN COMPANY

ACCOUNT SUMMARY

DATE 01/10/81 PAGE 1

PRIMARY ACCOUNT NO 161-2700231-6
STATEMENT PERIOD FROM 12/11/80 TO 01/10/81
BEGINNING BALANCE 800.00
ENDING BALANCE 389.41

Richard Miller
Harriet Miller
5341 Ashtree Lane
Cincinnati, Ohio 45211

DATE	TYPE	DESCRIPTION	DEPOSIT AMOUNT	WITHDRAWAL AMOUNT	BALANCE
80/12/15	DEP	Payroll Deposit	400.00		1,200.00
80/12/15	PAY	Marathon Oil Company		50.00	1,150.00
80/12/16	PAY	Lazarus Department Store		34.20	1,125.80
80/12/16	PAY	Ohio Power Company		150.00	965.80
80/12/18	CHK	101		20.00	945.80
80/12/19	DEP	Savings Deposit	50.00		995.80
80/12/19	W/D	Check Printing Charge		4.60	991.20
80/12/22	CHK	102		68.32	922.88
80/12/23	CHK	103		25.00	897.78
80/12/24	W/D	Preauthorized Mortgage Payment		375.00	522.88
80/12/16	W/D	Stop Pmt Hdlg Chg on Check #102		7.50	515.38
80/12/31	DEP	Payroll Deposit	400.00		915.38
80/12/31	DEP	Interest Deposit	8.62		924.00
81/01/02	CHK	104		20.00	904.00
81/01/02	CHK	105		75.00	829.00
81/01/05	DEP	Mail Deposit	150.00		979.00
81/01/07	CHK	106		44.59	934.41
81/01/08	CHK	107		25.00	909.41
81/01/08	PAY	Diamond Savings & Loan - Loan		350.00	559.41
81/01/09	PAY	VISA		150.00	409.41
81/01/10	CHK	108		20.00	389.41
		TOTALS	1,008.62	1,419.21	

Figure 17–5

It is important when you receive your bank statement and canceled checks that you **reconcile** your checkbook. This is a process of checking to be sure that your records and the bank's records agree. To reconcile your checkbook, you must bring yourself up to date on any action that the bank has taken. This would include service charges or interest paid. Write these amounts in your checkbook and either add or subtract the amount to the total (charges—subtract; interest earned—add). Then, bring the bank's total up to date on any checks you have written or deposits you have made that have not yet come through the bank's computer onto your statement. On a separate paper or on a form the bank might provide (usually on the back of the statement), begin with the bank's total for your account and add to it any deposits that you have made and subtract any checks that you have written that do not appear on the bank statement. This new total should match your new checkbook total. If it does, it is said that your checkbook **balanced**. If it does not, recheck your work to find your mistake. If you cannot get the two figures to match, telephone your bank as soon as possible and ask for help.

Example 1

Write the checks and the deposit slips required and record the following transactions on your check register:

March 15:   check to Phyllis Gordon for cosmetics, $12.57

March 18:   check to Riegle Electric for February bill, $35.22

March 19:   deposit two checks, $164.33 and $211.79, but receive $75.00 back in cash

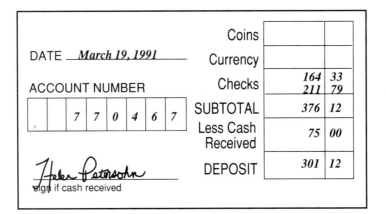

Helen Elaine Petersohn
1362 Danchel Boulevard
Brandon, IL 57233   No. 1274

March 15 19 _91_   3523-411
                          3788

PAY TO THE
ORDER OF ___Phyllis Gordon_____ $ 12 57/100

Twelve and 57/100 x x x x x x x x x x x x x x x x x x x x x x x x DOLLARS

Fosberg Bank, NA
Brandon, IL

5 7 2 4 3 6 2 - 7 7 0 4 6 7 - 1 2 7 4

Figure 17–6

Helen Elaine Petersohn
1362 Danchel Boulevard
Brandon, IL 57233   No. 1275

March 18 19 _91_   3523-411
                          3788

PAY TO THE
ORDER OF ___Rieale Electric_____ $ 35 22/100

Thirty-five and 22/100 x x x x x x x x x x x x x x x x x x x x x x DOLLARS

Fosberg Bank, NA
Brandon, IL

5 7 2 4 3 6 2 - 7 7 0 4 6 7 - 1 2 7 5

Figure 17–7

	Coins	
DATE _March 19, 1991_	Currency	
	Checks	164 | 33 211 | 79
ACCOUNT NUMBER	SUBTOTAL	376 | 12
⌐ | 7 | 7 | 0 | 4 | 6 | 7	Less Cash Received	75 | 00
sign if cash received	DEPOSIT	301 | 12

Figure 17–8

Number	Date	Description of Transaction	Payment		x	Deposit		Balance	
								512	79
1274	3/13/91	Phyllis Gordon (cosmetics)	12	57				500	22
1275	3/18/91	Riegle Electric (Feb. bill)	35	22				465	00
	3/19/91	DEPOSIT				301	12	766	12

Figure 17–9

Problems Complete the following checks and deposits.

1) Record all checks in problem 1 on this check register.

Number	Date	Description of Transaction	Payment	x	Deposit	Balance	
						1,531	78

Figure 17–10

Write a check to Barbara Langsley for a haircut ($8.50) on August 12, 1990.

Helen Elaine Petersohn
1362 Danchel Boulevard
Brandon, IL 57233

No. 1276

_____ 19 ____   3523-411
                    3788

PAY TO THE
ORDER OF _____ $ _____

_____ DOLLARS

Fosberg Bank, NA
Brandon, IL

5 7 2 4 3 6 2 - 7 7 0 4 6 7 - 1 2 7 6

Figure 17–11

Write a check to Hopewell Gas Company for July bill ($42.38) on August 12, 1990.

Helen Elaine Petersohn
1362 Danchel Boulevard
Brandon, IL 57233

No. 1277

_____ 19 ____   3523-411
                    3788

PAY TO THE
ORDER OF _____ $ _____

_____ DOLLARS

Fosberg Bank, NA
Brandon, IL

5 7 2 4 3 6 2 - 7 7 0 4 6 7 - 1 2 7 7

Write a check to First Church for an offering ($40.00) on August 14, 1990.

```
Helen Elaine Petersohn                              No. 1278
1362 Danchel Boulevard
Brandon, IL 57233                    _____ 19 ____   3523-411
                                                           3788

PAY TO THE
ORDER OF _____  $

_____ DOLLARS

Fosberg Bank, NA
Brandon, IL                          _____

5 7 2 4 3 6 2 - 7 7 0 4 6 7 - 1 2 7 8
```

Write a check to Farmer's Electric Company ($21.68) for July electric bill on August 16, 1990.

```
Helen Elaine Petersohn                              No. 1279
1362 Danchel Boulevard
Brandon, IL 57233                    _____ 19 ____   3523-411
                                                           3788

PAY TO THE
ORDER OF _____  $

_____ DOLLARS

Fosberg Bank, NA
Brandon, IL                          _____

5 7 2 4 3 6 2 - 7 7 0 4 6 7 - 1 2 7 9
```

Make a deposit of $4.25 in coins and a check for $350.77 on August 16, 1990.

```
                Deposit Slip        Coins    |      |
                                             |------|------|
DATE _____           Currency |      |      |
                                             |------|------|
ACCOUNT NUMBER                        Checks  |      |      |
                                             |------|------|
| | | | | | | | |                  SUBTOTAL  |      |      |
                                             |------|------|
                                   Less Cash  |      |      |
                                   Received   |      |      |
                                             |------|------|
_____               DEPOSIT  |      |      |
sign if cash received
```

Figure 17–12

Write a check to Betsy Graham for a decorated cake ($15.50) on August 20, 1990.

```
Helen Elaine Petersohn                              No. 1280
1362 Danchel Boulevard
Brandon, IL 57233                    _____ 19 ____   3523-411
                                                           3788

PAY TO THE
ORDER OF _____  $

_____ DOLLARS

Fosberg Bank, NA
Brandon, IL                          _____

5 7 2 4 3 6 2 - 7 7 0 4 6 7 - 1 2 8 0
```

Write a check to Haven Hardware for car parts ($45.22) on August 25, 1990.

Helen Elaine Petersohn		No. 1281
1362 Danchel Boulevard		
Brandon, IL 57233	_____ 19 ___	3523-411
		3788
PAY TO THE		$
ORDER OF _____		
_____ DOLLARS		
Fosberg Bank, NA		
Brandon, IL		
		_____
5724362-770467-1281		

Write a check to Jacob's Jewelry Store ($50.00) for a deposit on a ring on August 27, 1990.

Helen Elaine Petersohn		No. 1282
1362 Danchel Boulevard		
Brandon, IL 57233	_____ 19 ___	3523-411
		3788
PAY TO THE		$
ORDER OF _____		
_____ DOLLARS		
Fosberg Bank, NA		
Brandon, IL		
		_____
5724362-770467-1282		

2) Record all checks in problem 2 on this check register.

Number	Date	Description of Transaction	Payment	x	Deposit	Balance	
						1,531	78

Write a check to Parson's Grocery for groceries ($56.70) on April 13, 1990.

Helen Elaine Petersohn		No. 1283
1362 Danchel Boulevard		
Brandon, IL 57233	_____ 19 ___	3523-411
		3788
PAY TO THE		$
ORDER OF _____		
_____ DOLLARS		
Fosberg Bank, NA		
Brandon, IL		
		_____
5724362-770467-1283		

Write a check to Brighton Land Company for July rent ($240.00) on April 14, 1990.

Helen Elaine Petersohn
1362 Danchel Boulevard
Brandon, IL 57233

No. 1284

_____ 19 _____

3523-411
3788

PAY TO THE
ORDER OF _____ $ _____

_____ DOLLARS

Fosberg Bank, NA
Brandon, IL

_____

5 7 2 4 3 6 2 - 7 7 0 4 6 7 - 1 2 8 4

Write a check to Red Cross for a donation ($30.00) on April 15, 1990.

Helen Elaine Petersohn
1362 Danchel Boulevard
Brandon, IL 57233

No. 1285

_____ 19 _____

3523-411
3788

PAY TO THE
ORDER OF _____ $ _____

_____ DOLLARS

Fosberg Bank, NA
Brandon, IL

_____

5 7 2 4 3 6 2 - 7 7 0 4 6 7 - 1 2 8 5

Write a check to Blanchard Electric Company ($27.83) for March electric bill on April 20, 1990.

Helen Elaine Petersohn
1362 Danchel Boulevard
Brandon, IL 57233

No. 1286

_____ 19 _____

3523-411
3788

PAY TO THE
ORDER OF _____ $ _____

_____ DOLLARS

Fosberg Bank, NA
Brandon, IL

_____

5 7 2 4 3 6 2 - 7 7 0 4 6 7 - 1 2 8 6

Make a deposit of $100.00 in currency and 2 checks, for $35.00 and $190.35, on April 21, 1990.

Deposit Slip

DATE _____

ACCOUNT NUMBER

_____
sign if cash received

Coins		
Currency		
Checks		
SUBTOTAL		
Less Cash Received		
DEPOSIT		

Write a check to Karen Hanover for babysitting ($5.50) on April 20, 1990.

```
┌─────────────────────────────────────────────────────────────┐
│  Helen Elaine Petersohn                          No. 1287     │
│  1362 Danchel Boulevard                                       │
│  Brandon, IL 57233              _____ 19 ___    3523-411   │
│                                                      3788      │
│                                                               │
│  PAY TO THE                                      ┌──────────┐ │
│  ORDER OF _____   │ $        │ │
│                                                  └──────────┘ │
│  _____ DOLLARS   │
│                                                               │
│  Fosberg Bank, NA                                             │
│  Brandon, IL                    _____ │
│                                                               │
│  5 7 2 4 3 6 2 - 7 7 0 4 6 7 - 1 2 8 7                        │
└─────────────────────────────────────────────────────────────┘
```

Write a check to Gross Electric for new lamps ($48.26) on April 25, 1990.

```
┌─────────────────────────────────────────────────────────────┐
│  Helen Elaine Petersohn                          No. 1288     │
│  1362 Danchel Boulevard                                       │
│  Brandon, IL 57233              _____ 19 ___    3523-411   │
│                                                      3788      │
│                                                               │
│  PAY TO THE                                      ┌──────────┐ │
│  ORDER OF _____   │ $        │ │
│                                                  └──────────┘ │
│  _____ DOLLARS   │
│                                                               │
│  Fosberg Bank, NA                                             │
│  Brandon, IL                    _____ │
│                                                               │
│  5 7 2 4 3 6 2 - 7 7 0 4 6 7 - 1 2 8 8                        │
└─────────────────────────────────────────────────────────────┘
```

Write a check to Gonago City Bank ($250.00) for April car payment on April 25, 1990.

```
┌─────────────────────────────────────────────────────────────┐
│  Helen Elaine Petersohn                          No. 1289     │
│  1362 Danchel Boulevard                                       │
│  Brandon, IL 57233              _____ 19 ___    3523-411   │
│                                                      3788      │
│                                                               │
│  PAY TO THE                                      ┌──────────┐ │
│  ORDER OF _____   │ $        │ │
│                                                  └──────────┘ │
│  _____ DOLLARS   │
│                                                               │
│  Fosberg Bank, NA                                             │
│  Brandon, IL                    _____ │
│                                                               │
│  5 7 2 4 3 6 2 - 7 7 0 4 6 7 - 1 2 8 9                        │
└─────────────────────────────────────────────────────────────┘
```

Personal Savings

Having a savings account can be a very wise way to save money when you have a steady income. A savings account usually requires that you keep a minimum balance in the account, and interest is earned on the money that is deposited. The advantage to a savings account over other types of saving—certificates of deposit, U.S. savings bonds, mutual funds, Christmas club accounts, and so forth—is that your money is readily available to you. There is no penalty for withdrawal of funds nor are there any requirements on when deposits can be made.

To open a savings account, you must first select a bank. Banks vary in the amount of interest they will pay on various savings accounts—usually depending on how much money you are planning to keep in the account (what your minimum balance will be). Once you have selected a bank in which to open your account, you must go to that bank and fill out the appropriate application and forms for opening a savings account. You must deposit a certain amount of money (at least the minimum specified by that bank). You will receive a passbook (similar to a check register) when the account has been opened. This is a book in which you record all transactions (deposits, withdrawals, and interest earned) that happen with your account.

You can make deposits or withdrawals whenever you wish in your savings account. A deposit is completed like a checking account deposit. A withdrawal of money can be made by filling out a withdrawal form at the bank. The withdrawal form requires the following information:

- date
- account number
- amount in numbers and in words
- your signature

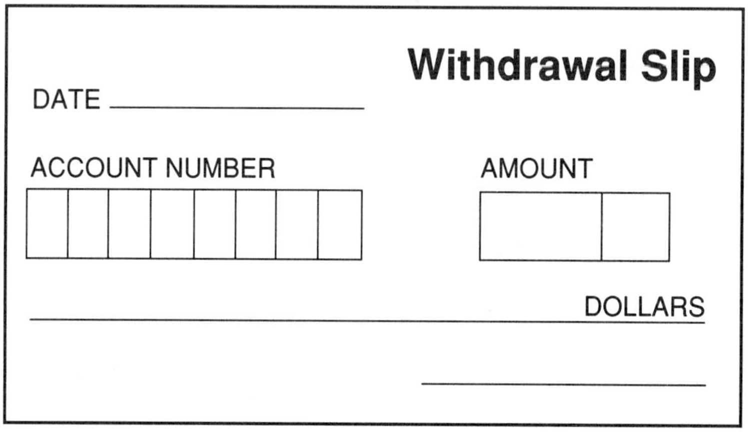

```
┌─────────────────────────────────────────────────────────┐
│                              Withdrawal Slip              │
│                                                           │
│   DATE _____                                    │
│                                                           │
│   ACCOUNT NUMBER                   AMOUNT                 │
│   ┌──┬──┬──┬──┬──┬──┬──┐          ┌────┬────┐            │
│   └──┴──┴──┴──┴──┴──┴──┘          └────┴────┘            │
│                                                           │
│   _____     DOLLARS           │
│                                                           │
│                               _____        │
└─────────────────────────────────────────────────────────┘
```

Figure 17–13

Once your account has been opened, you will receive statements from the bank every month, quarter (three months), or other time period agreed upon when you open your account. This statement is similar to a checking account statement. You will need to record the interest you have received in the passbook. Then you should check to see if the bank balance agrees with your own record in the passbook.

Problems   Complete the following deposit slips and withdrawal slips for Gina Evans (acct. no. 46–16333).

1) Deposit $.75 in coins, $25.00 in currency, and $153.67 in checks. No cash received.

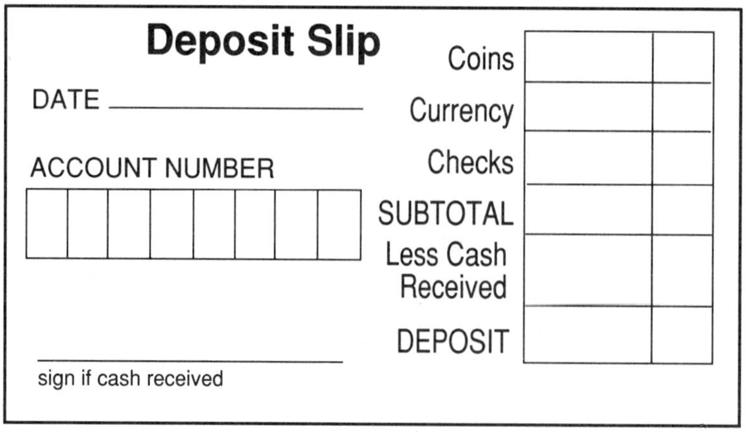

```
┌──────────────────────────────────────────────────────┐
│      Deposit Slip        Coins    │        │         │
│                                   ├────────┼─────────┤
│   DATE _____    Currency │        │         │
│                                   ├────────┼─────────┤
│   ACCOUNT NUMBER         Checks   │        │         │
│   ┌──┬──┬──┬──┬──┬──┬──┐          ├────────┼─────────┤
│   └──┴──┴──┴──┴──┴──┴──┘ SUBTOTAL │        │         │
│                          Less Cash├────────┼─────────┤
│                          Received │        │         │
│   _____          ├────────┼─────────┤
│   sign if cash received  DEPOSIT  │        │         │
│                                   └────────┴─────────┘
└──────────────────────────────────────────────────────┘
```

2) Withdraw $160.00.

```
+--------------------------------------------------+
|                                                  |
|                          Withdrawal Slip         |
|  DATE _____                           |
|                                                  |
|  ACCOUNT NUMBER              AMOUNT               |
|  +-+-+-+-+-+-+-+-+          +------+------+       |
|  | | | | | | | | |         |      |      |       |
|  +-+-+-+-+-+-+-+-+          +------+------+       |
|                                                  |
|                                   DOLLARS        |
|  _____          |
|                                                  |
|                              _____   |
+--------------------------------------------------+
```

3) Deposit 2 checks, for $165.99 and $274.83. Receive $45.00 back in cash.

```
+------------------------------------------------------+
|   Deposit Slip      Coins    +--------+--------+      |
|                              |        |        |      |
|  DATE _____ Currency+--------+--------+      |
|                              |        |        |      |
|  ACCOUNT NUMBER      Checks  +--------+--------+      |
|  +-+-+-+-+-+-+-+-+           |        |        |      |
|  | | | | | | | | |  SUBTOTAL +--------+--------+      |
|  +-+-+-+-+-+-+-+-+  Less Cash|        |        |      |
|                     Received +--------+--------+      |
|  _____ DEPOSIT |        |        |      |
|  sign if cash received       +--------+--------+      |
+------------------------------------------------------+
```

4) Make a deposit of $100.00 in currency, $2.78 in coins, and two checks, $178.55 and $273.71.

```
+------------------------------------------------------+
|   Deposit Slip      Coins    +--------+--------+      |
|                              |        |        |      |
|  DATE _____ Currency+--------+--------+      |
|                              |        |        |      |
|  ACCOUNT NUMBER      Checks  +--------+--------+      |
|  +-+-+-+-+-+-+-+-+           |        |        |      |
|  | | | | | | | | |  SUBTOTAL +--------+--------+      |
|  +-+-+-+-+-+-+-+-+  Less Cash|        |        |      |
|                     Received +--------+--------+      |
|  _____ DEPOSIT |        |        |      |
|  sign if cash received       +--------+--------+      |
+------------------------------------------------------+
```

5) Withdraw $271.88.

Withdrawal Slip

DATE _____

ACCOUNT NUMBER

AMOUNT

DOLLARS

6) Deposit three checks: $512.22, $173.27, and $25.00. Receive $40.75 back in cash.

Deposit Slip

DATE _____

ACCOUNT NUMBER

sign if cash received

Coins
Currency
Checks
SUBTOTAL
Less Cash Received
DEPOSIT

7) Withdraw $176.00.

Withdrawal Slip

DATE _____

ACCOUNT NUMBER

AMOUNT

DOLLARS

8) Withdraw $1,747.79.

Withdrawal Slip

DATE _____

ACCOUNT NUMBER   AMOUNT

DOLLARS

_____

9) Deposit two checks, $572.44 and $1,376.73, and receive $160.00 back in cash.

Deposit Slip

DATE _____

ACCOUNT NUMBER

sign if cash received

	Coins		
	Currency		
	Checks		
	SUBTOTAL		
	Less Cash Received		
	DEPOSIT		

10) Deposit 2 rolls of quarters (worth $10 each), 3 rolls of nickels (worth $2 each), loose coins of $1.57, currency of $675, and a check for $383.76.

Deposit Slip

DATE _____

ACCOUNT NUMBER

sign if cash received

	Coins		
	Currency		
	Checks		
	SUBTOTAL		
	Less Cash Received		
	DEPOSIT		

Credit

Using credit is the same as borrowing money for an item. When you use credit, you are required to pay interest on the money borrowed and return the money either in a lump sum or in equal monthly payments. Credit should be used with great caution because it often gives "false security"—the feeling that you can afford something that you really cannot afford.

Types of Credit

There are several different types of credit. Borrowing for a car, a personal loan, a house loan, loans for college, and credit cards are all examples of credit. In this section we will focus on the use of credit cards.

There are different types of credit cards: major credit cards, credit cards from a specific store, gasoline cards, telephone credit cards, and so forth. The major credit cards are the most difficult to obtain since they usually extend the largest amount of credit (the limit that you are allowed to borrow). These cards almost always charge a yearly fee to use the card and charge interest every month on the unpaid balance. Credit cards from a specific store do not normally charge annual fees, but they do charge interest every month. They usually do not extend as much credit. Gasoline cards often do not charge an annual fee and they normally require you to pay the entire balance at the end of the month. When the balance is not paid, they charge interest on the unpaid balance. Telephone credit cards are similar to gasoline credit cards.

Obtaining Credit

Obtaining credit can be difficult for young people who have never had credit before. It can also be difficult to obtain if you have a bad **credit rating**. Your credit rating is a record of how you have handled credit in the past—how promptly your debts have been paid, how reliable you have been, the extend of your debt, and other factors.

To apply for credit, you are required to fill out an application. Credit applications usually ask for the following information:
- name, address, phone number
- name, address, phone number of relative not living with you (for reference)
- occupation, place of employment, and salary
- work history
- banking information (location, type, and amount of money in accounts held)
- credit history
- references

Using Credit

Credit should be used wisely. Using credit is often easier than paying by cash or check, but it should not be abused. The benefits of credit include carrying smaller amounts of cash and knowing that your card will be accepted in a different region of the country. When you are traveling, a credit card enables you to reserve hotel rooms in advance, make long distance phone calls with ease, and pay for meals, gasoline, and other necessities.

If you use credit, it is wise to follow these guidelines:
- do not charge more than you can afford to pay for each month
- pay for all credit card bills *in full* the month you receive them
- do not charge items on impulse that you do not need

If you follow these guidelines, you will not have a problem with abusing credit. These suggestions are a strict approach to credit, but they will help you to avoid many of the problems that arise with credit abuse.

Paying the Bills That Are Incurred

When you charge items on a credit card, you are sent a monthly statement of that account. The statement usually shows a record of the purchases and the payments made during the previous month. It indicates your new balance and the minimum amount that you are required to pay. It is best to pay the entire balance at the end of each month but you are required to pay only the minimum amount. The rest of the charge amount will be carried forward to the next month and interest will be assessed.

To pay these bills, you simply write a check for the amount you have decided to pay. There is usually a part of the bill that they request you to return to them with your payment. Detach that part of the bill, fill out any information requested, and put that and your check in an envelope and send it to the company. Be sure to record your check on your check register.

Problems   Answer the following questions.

1)   List three different types of credit. _____
_____

2)   What are three guidelines to use credit wisely? _____
_____

3)   What is "false security"? _____
_____

Utility Bills

Under the category of utility bills, we will look at all monthly expenses for running a home. These include **household expenses**, such as heating (gas, oil, electric, etc.), electricity, telephone, cable television, garbage collection, water, and others. We will also include **fixed expenses** such as taxes and home insurance.

If you own your home, you are responsible for all of these expenses and your mortgage payment. If you rent, your landlord may provide some of these services—water, garbage, taxes, insurance for the structure, and so forth. It is wise for renters to look into property insurance for their possessions, however, because the landlord's insurance policy usually will not cover personal possessions. Renter's insurance for possessions is fairly inexpensive and can be a wise investment in case of fire or theft.

Most household expenses are paid monthly. Since the amounts can fluctuate greatly from month to month (due to the weather, greater use, etc.), some companies offer a budget plan. A **budget plan** is a method of paying the same amount each month based on the average amount due. For example, if the electric bills for the previous year ran $45, $40, $40, $35, $55, $50, $50, $45, $55, $60, $65, and $60 for the twelve months, an average amount would be found ($50) and that amount would be charged every month for the following year. If in the final month too much or not enough had been paid to cover the actual costs, an adjustment would be made.

The fixed expenses are usually paid once or twice a year. Since these are usually large amounts, it is wise to save each month for the expense so that when the bill comes, you will have the money to pay it. For example, if the yearly insurance cost is $372, each month's portion is $31 ($372 ÷ 12 = $31). You can save $31 each month, and when the bill comes at the end of the 12 months, you will have enough money saved to pay it.

To know how much to set aside to pay fixed expenses

1. Determine how many months the expense covers (annual = 12 months, twice a year = 6 months, etc.).

2. Divide the amount of the bill by the number of months.

Notice in this sample utility bill the following features:

- previous balance
- payment from previous month
- current balance
- due date
- amount due

Northern Illinois Gas

Date prepared: August 12, 1991

Gas used at: 153 Hope Drive
Your Acct. No: 163167428383

Your meter read 878.7 on Aug. 1
Your meter read 875.3 on July 1
Your usage was 3.4 thousand cubic feet. Next meter reading will be about September 1.

Monthly Service Charge $ 2.17
  3.4 MCF at $4.5   15.30
                          17.47

Account Summary

PREVIOUS BALANCE	$ 13.35
PAYMENT	13.35
CURRENT BILL	17.47
CURRENT BALANCE	17.47

PLEASE PAY: $17.47
        BY: Aug. 15, 1991

Figure 17–14

Problems   Read the following utility bills and write a check to the company for the appropriate amount. Date your checks according to the date the bill is due.

1)

Northern Illinois Gas

Date prepared: August 12, 1991

Gas used at: 153 Hope Drive
Your Acct. No: 163167428383

Your meter read 878.7 on Aug. 1
Your meter read 875.3 on July 1
Your usage was 3.4 thousand cubic feet. Next meter reading will be about September 1.

Monthly Service Charge $ 2.17
  3.4 MCF at $4.5   15.30
                          17.47

Account Summary

PREVIOUS BALANCE	$ 13.35
PAYMENT	13.35
CURRENT BILL	17.47
CURRENT BALANCE	17.47

PLEASE PAY: $17.47
        BY: Aug. 15, 1991

Helen Elaine Petersohn
1362 Danchel Boulevard
Brandon, IL 57233

No. 1290

_____ 19 ___   3523-411
                     3788

PAY TO THE
ORDER OF _____ $ _____

_____ DOLLARS

Fosberg Bank, NA
Brandon, IL

5724362-770467-1290

2)

ACCOUNT NUMBER: 274-7382-22   SUBSCRIBER NAME: Jennifer Myer

DUE DATE: March 9, 1991

DATE		DESCRIPTION	AMOUNT	BALANCE
From	To			
3/1		Previous balance	17.95	
3/1	3/31	Cable television SVC	17.95	
2/7		Prompt payment	17.95	
		AMOUNT DUE ————————		**17.95**

Make payment to: **ANDERSON CABLE**
1743 VISTA BLVD.
BRANDON, IL 57233

Figure 17–15

Helen Elaine Petersohn
1362 Danchel Boulevard
Brandon, IL 57233

No. 1291

_____ 19 ____   3523-411
                        3788

PAY TO THE
ORDER OF _____ $ _____

_____ DOLLARS

Fosberg Bank, NA
Brandon, IL

5 7 2 4 3 6 2 - 7 7 0 4 6 7 - 1 2 9 1

---

3)

Allied Telephone Company   Pg. 1
March 1, 1991

Business Office: 267-7113   **Svc. for: 267-4513**

Last Month's Charges ———————————————— 28.27
Payment ——————— 2/6/91 ———————————— 28.27
Adjustments ——————————————————————— –0–
Amount Due Previous Bill———————————— .00
TOTAL CURRENT CHARGES ————————————— 32.78

TOTAL AMOUNT DUE ———————————————— **32.78**

DUE BY: 3/10/91

Local Service—Feb. 1 to March 1   27.55
Other Service—Feb. 1 to March 1   3.12
Taxes   2.11
                                    32.78

Figure 17–16

Helen Elaine Petersohn
1362 Danchel Boulevard
Brandon, IL 57233

No. 1292

_____ 19 ____   3523-411
                        3788

PAY TO THE
ORDER OF _____ $ _____

_____ DOLLARS

Fosberg Bank, NA
Brandon, IL

5 7 2 4 3 6 2 - 7 7 0 4 6 7 - 1 2 9 2

4)

```
                  MIDWEST POWER COMPANY
                      178 MOLAIR DRIVE
                      BRANDON, IL 57233

   SVC. NAME  Jennifer Myers              ACCT. NO. 472747289
              1674 Raymond Dr. West
              Brandon, IL 57233


   FUEL RATE PER KWH—1.881690

   SERVICE PERIOD          METER READINGS          KWH USED
     from     to         previous    present      THIS MONTH
     2/5      3/5          52687      53204           517


   DESCRIPTION                          AMOUNT
   Balance as of Last Billing Date       36.26
   Payment 2/10 Thank You                36.26
        Previous Balance                   .00
   517 KWH Used This Period              38.17
        TOTAL                            38.17


   $1.27 Average Cost a Day
   Last Pay Date—March 10
   Pay This Amount  $38.17
```

Figure 17–17

```
   Helen Elaine Petersohn                         No. 1293
   1362 Danchel Boulevard
   Brandon, IL 57233            _____ 19 ___    3523-411
                                                      3788

   PAY TO THE                                 $
   ORDER OF _____

   _____ DOLLARS

   Fosberg Bank, NA
   Brandon, IL                  _____

   5 7 2 4 3 6 2 - 7 7 0 4 6 7 - 1 2 9 3
```

5)

```
   AUSTIN ENTERPRISES               NO. 1372-66
   173 RANDALL COURT
   BRANDON, IL 57233

   For rent at residence,          March rent: $370.00
        1674 Raymond Drive West    Due by: 3/15/91
        Brandon, IL 57233
```

Figure 17–18

```
   Helen Elaine Petersohn                         No. 1294
   1362 Danchel Boulevard
   Brandon, IL 57233            _____ 19 ___    3523-411
                                                      3788

   PAY TO THE                                 $
   ORDER OF _____

   _____ DOLLARS

   Fosberg Bank, NA
   Brandon, IL                  _____

   5 7 2 4 3 6 2 - 7 7 0 4 6 7 - 1 2 9 4
```

Purchasing a Car

When purchasing a car, you have two choices. You can buy a new car or a used one. **New cars** are ones that have had no other owner besides the dealer, whether they are the current year's model or a year old. **Used cars** are ones that have had at least one previous owner. You may buy a used car from that previous owner or from a used car dealer.

Cars are identified by the company that makes them. This identification is called the **make** of the car. For example, a car made by Ford Motor Company is known as a Ford—this is the make. Then, the manufacturer also identifies the car with another name. This is called the **model**. For example, you might go to a Ford dealer and look at several Ford automobiles—the Escort, the Mustang, and so forth. Then, according to the body type and style the car might be further identified as a **two-door, four-door, hatchback, coupe, sedan, station wagon**, and the like.

When you purchase a new car, it is important to shop around to select the make and model that best meets your needs and your budget. Then, once you've selected the type of car that you are interested in, it is helpful if you can visit more than one dealer to compare prices, service, and selection.

There are different qualities available in used cars. Because of this, you need to be even more aware of what type of car best meets your needs and your budget. You might find that your needs could be met by various makes and models, and it will be your goal to get the best car for your money.

When purchasing a used car, you need to find out if there is a guarantee on any or all of the vehicle. You will also want to know how many miles the car has been driven (this is found on the **odometer**), the condition of the battery and tires, and general performance of the engine. If you do not have a good understanding of cars, it is wise to take a friend or relative along with you when you shop.

On a new car, the considerations are different. You will usually be given warranty information, and it is important to understand what guarantee comes with the car and what guarantee features would have to be purchased separately.

In both cases, you will want to test drive the vehicle. Use this drive to test all of the accessories (electric locks, windows, radio, etc.—especially in a used vehicle). You will also want to consider the appearance and the comfort of the vehicle.

When you have selected the vehicle that you wish to buy, you are required to either pay the total amount or make some kind of a down payment. A **down payment** is a partial payment on the vehicle, and it can be in the form of cash or check. Most car dealers will allow you to use a trade-in allowance as part (or all) of your down payment. A **trade-in allowance** is a cash allowance the dealer gives you when you trade your current vehicle in to him in exchange for the newer one. If you make a down payment, you will have to get a loan for the remaining amount owed.

A car loan is a type of installment loan. An **installment loan** is a bank loan where you borrow a certain amount and then make monthly payments to repay the principal and the interest. You are usually required to have some collateral on an installment loan, and in this case, the banks use the car as collateral. **Collateral** is property of significant value that the bank could receive if you refuse to pay the loan. Since the bank uses the car as collateral, they will keep the title (certificate of ownership) to the car until you have paid for it entirely. Then, they will give the title to you and the car is officially yours.

In most cases, to get a bank loan, you must have some credit established or you must have a cosigner for your loan. A **cosigner** is someone who is willing to bear the responsibility of the loan if you refuse to pay it. Being a cosigner should not be taken lightly—it is a large financial and legal responsibility. It is far better for you (and your cosigner) if you have established credit on your own first and do not need a cosigner for the loan. Credit can be established by applying for (and receiving) credit cards or other loans. To have a good credit rating, it is important to pay all credit card bills or loan payments promptly, since this information is available to other banks or institutions when you try to receive more credit.

Once the bank agrees to grant you a car loan, you will make monthly payments. Out of this payment, a certain part goes toward paying off the principal (amount borrowed) and a certain part goes toward paying off the interest (fee charged for the use of the money). At the beginning of a loan, the majority of the payment goes toward paying off the interest. The exact amount of the payment that is used to pay off the principal can be figured using a simple process.

To figure the interest, we will use the formula:

$$I = P \times R \times T$$

This stands for Interest = Principal × Rate × Time

To find the amount of the payment going toward interest, we will multiply the principal (amount still due) times the rate (the interest rate—a percent) times the time (1 month—1 out of 12 or $\frac{1}{12}$). This tells us what part of the payment is going toward the interest. We can then subtract that amount from the monthly payment to determine how much of the principal has been paid.

Example 1

The Johnsons recently bought a new car. They obtained a loan for $11,579 at 15% interest. Their monthly payment is $195. How much of the first payment will be put on the interest?

1. Find the amount going toward interest. $I = P \times R \times T$

   $I = \$11,579 \times 15\% \times \frac{1}{12}$

   change to decimal   multiply by 1, divide by 12

   $= 11,579 \times .15 \div 12$

   $= \$144.7375$ rounds to $144.74

2. Find the amount left from the payment to go toward principal.

   Principal = Monthly payment – interest

   $= \$195.00 - 144.74$

   Principal = $50.26

3. $50.26 will be put on the principal. We could also find the new balance. We owed $11,579 and have paid $50.26. Our new balance is: $11,579 - 50.26 = \$11,528.74$.

Problems   Answer the following questions about car loans.

1) Carolyn Wagner has a loan on her new minivan. She currently owes $9,572.78. Her monthly payment is $215.75, and her loan has an interest rate of $17\frac{1}{4}\%$. What is the new balance on her loan after her next payment?   _____

2) Kirk Von Kemp recently bought a used two-door sedan. He obtained a loan for $6,000 at 18.5%. His monthly payment is $255.59.
   What is the amount of the first payment that will go toward interest?   _____
   What is the amount of the first payment that will go toward principal?   _____
   How much does he owe on the loan after the first payment?   _____

3) Delores Hafner pays $215.90 a month for her car payment. She owes $7,500 on the car at an interest rate of $15\frac{1}{4}\%$. What is the new balance on her loan after her next payment?   _____

4) Barb Greensley just bought a Pontorolla Sedan for $15,527.70. She paid $3,000 down and borrowed the remaining amount from the Land Bank of Wilsonville at 16%. She makes monthly payments of $264.
What is the amount of the first payment that will go toward interest?   _____
What is the amount that will go toward principal?   _____
How much does she owe on the loan after the first payment?   _____

5) Alexandria Higgins just bought a sports car for $18,895.95. She used her old car as a down payment (they gave her $4000 credit for it). She borrowed the remaining amount from her bank at 15.5%. She makes monthly payments of $217.78. What does she owe on the loan after the first payment?   _____

6) Ray Grove owes $3648.88 on his Pridemont station wagon. He makes $155.18 car payments and has borrowed the money at 11.5%.
How much of his next payment will go toward interest?   _____
How much will go toward principal?   _____
What will his new balance be?   _____

7) August Summerfield wants to pay off the loan on her Jasper minivan next month. This month she makes a $248.00 car payment. She owes $1579.12, and her loan is at 12% interest. How much will she have to pay next month to pay the entire balance of the loan?   _____

8) Tracy Reese needs to make a 15% down payment on the new car she has selected. If it costs $16,840, what will the required down payment be?   _____

9) Natasha Vickers picks out a used car that she wishes to buy. It costs $6784.95, and she must pay 25% down. She will borrow the rest from Highlandville Savings and Loan.
What will the required down payment be?   _____
How much will she borrow from the bank?   _____

10) Cathy Crossly wants to buy a truck that costs $15,648.90. She will pay 20% down and borrow the rest from her credit union. Her monthly payments will be $227.99 at $12\frac{1}{4}\%$ interest.
How much of her first payment will go toward interest?   _____
How much will go toward principal?   _____
What will her new balance be?   _____

Housing

There are several choices available in the housing market. When you wish to live out on your own, you can choose between renting or buying. Places available for rent include apartments, town houses, efficiencies (one- or two-room apartments), mobile homes, and houses. Places available for purchase include condominiums, mobile homes, and houses. There is a variety of other options available in certain areas with the agreement of the owner or landlord. You can decide to live alone or share the expenses with another individual.

When you consider housing, it is important to understand what features are included in your monthly payment. For example, some rental agreements include payment of your utilities. You could expect this to have a higher monthly payment than if you are required to pay for your own utility bills. Some agreements include certain features—perhaps the water bill or garbage collection—and not others. Some agreements do not cover any extra features. It is also important to find out how much commitment is involved in the agreement. For example, when renting, you may be asked to sign a one-year lease. This obligates you to pay rent for one year. Or in a purchasing situation, you obviously will be bound to your agreement until you can sell to someone else.

Figure 17–19

There are extra expenses that people often overlook when selecting housing. Do you have to pay a **deposit** (a one-time payment that covers any damage you might do that is refunded when you leave the place in good condition)? Do you have to make a down payment if buying? Is there a **monthly maintenance fee** (a fee collected in some condominium or mobile home situations that pays for grounds upkeep, security, outdoor lighting, etc.)? Is the place furnished or unfurnished? Are drapes and/or carpeting provided? What about insurance, taxes, and so forth? Be wise in your selection of housing, and be sure that you can afford it.

To choose housing, you must first decide whether you will rent or buy. Use affordability as a factor, as well as how long you plan to be in the area, how well you know the area, and so forth. Next, determine how much you can afford to spend on housing (this will be covered in the Monthly Budget section). Decide what area of town you want to be in or whether you would prefer a more rural area. Then, begin looking in the classified ads in your local newspaper. Circle options that you would like to consider further and call the realtor or landlord. Ask questions about each place, its features, extra costs involved, the commitment, and the like. Make arrangements to see the places that interest you and then make your decision.

Monthly Budget

The most successful way of handling your money is to use a budget. A **budget** is simply a plan for spending and saving your money. It is not a set of restrictions that you have no control over, but a guideline to help you get the most out of your money.

There are several ways to set up a budget. We will be looking at the percent method, which allocates a certain percent of your income for housing, a certain percent for transportation (car payment, insurance, taxes, repairs, etc.), a certain percent for clothing, and so forth. This helps you to have a balanced spending plan. It can be adjusted to meet your own specific needs. For example, perhaps nice housing is really important to you. You may adjust the percent allotted for housing to a higher rate, then spend less in some other, less-important area.

The percents recommended are as follows:

Housing	32%	Entertainment/Recreation	7%
Food	15%	Clothing	5%
Car	15%	Savings	5%
Insurance	5%	Miscellaneous	6%
Debt	5%		

These percents add up to 95%. The remaining 5% may be allotted to a category where there is greater need or to an additional category not considered here. To be successful financially, your budget should never exceed 100% of your net spendable income. (Use a four-week month whenever weekly income is given. In months with five paydays, the extra check can be used for vacations, Christmas, or other special projects.)

To determine how much you can afford in each of these areas, multiply your monthly net spendable income by the appropriate percent. Let's look at a sample budget.

Example 1

Carole Havner earns $225/weekly (net pay). Determine how much she can afford to spend in each of the above budget categories.

1. Find her monthly net income ($225 × 4 = $900).

2. Multiply $900 by each percent.

Housing	32% × 900 =	$288	Entertainment	7% × 900 =	63
Food	15% × 900 =	135	Clothing	5% × 900 =	45
Car	15% × 900 =	135	Savings	5% × 900 =	45
Insurance	5% × 900 =	45	Miscellaneous	6% × 900 =	54
Debt	5% × 900 =	45	Left over	5% × 900 =	45

Problems   Determine how much each individual can afford to spend in each of the budget categories.

1) Alicia Grace earns $15,770/yearly net pay.

   a. What is her monthly net pay?   _____

   b. What is her budget amount in:

Housing	_____	Entertainment/Recreation	_____
Food	_____	Clothing	_____
Car	_____	Savings	_____
Insurance	_____	Miscellaneous	_____
Debt	_____	Left over	_____

2) George DeNisco earns $1662/monthly net pay.

   a. What is his monthly net pay?   _____

   b. What is his budget amount in:

Housing	_____	Entertainment/Recreation	_____
Food	_____	Clothing	_____
Car	_____	Savings	_____
Insurance	_____	Miscellaneous	_____
Debt	_____	Left over	_____

3) Charlotte Morgan earns $590/weekly net pay.

   a. What is her monthly net pay?   _____

   b. What is her budget amount in:

Housing	_____	Entertainment/Recreation	_____
Food	_____	Clothing	_____
Car	_____	Savings	_____
Insurance	_____	Miscellaneous	_____
Debt	_____	Left over	_____

4) Harold Bonfilliet earns $27,675/yearly gross pay. Twenty-seven percent of his check is taken out in taxes and other deductions (hint: find his yearly **net** pay first).

    a. What is his monthly net pay?   _____

    b. What is his budget amount in:

Housing	_____	Entertainment/Recreation	_____
Food	_____	Clothing	_____
Car	_____	Savings	_____
Insurance	_____	Miscellaneous	_____
Debt	_____	Left over	_____

5) Austin Long earns $5178/monthly net pay.

    a. What is his monthly net pay?   _____

    b. What is his budget amount in:

Housing	_____	Entertainment/Recreation	_____
Food	_____	Clothing	_____
Car	_____	Savings	_____
Insurance	_____	Miscellaneous	_____
Debt	_____	Left over	_____

6) Justine Graham-McGuire earns $37,881/yearly net pay.

    a. What is her monthly net pay?   _____

    b. What is her budget amount in:

Housing	_____	Entertainment/Recreation	_____
Food	_____	Clothing	_____
Car	_____	Savings	_____
Insurance	_____	Miscellaneous	_____
Debt	_____	Left over	_____

7) Connie Hawasako earns $23,404/yearly net pay.

    a. What is her monthly net pay?   _____

    b. What is her budget amount in:

Housing	_____	Entertainment/Recreation	_____
Food	_____	Clothing	_____
Car	_____	Savings	_____
Insurance	_____	Miscellaneous	_____
Debt	_____	Left over	_____

8) Ray Kildear earns $292.88/weekly net pay.

    a. What is his monthly net pay?   _____

    b. What is his budget amount in:

Housing	_____	Entertainment/Recreation	_____
Food	_____	Clothing	_____
Car	_____	Savings	_____
Insurance	_____	Miscellaneous	_____
Debt	_____	Left over	_____

9) Paul Chung earns $792/monthly net pay.

    a. What is his monthly net pay?   _____

    b. What is his budget amount in:

Housing	_____	Entertainment/Recreation	_____
Food	_____	Clothing	_____
Car	_____	Savings	_____
Insurance	_____	Miscellaneous	_____
Debt	_____	Left over	_____

10) Carissa Rostenkiak earns $92,902/yearly net pay.

    a. What is her monthly net pay?   _____

    b. What is her budget amount in:

Housing	_____	Entertainment/Recreation	_____
Food	_____	Clothing	_____
Car	_____	Savings	_____
Insurance	_____	Miscellaneous	_____
Debt	_____	Left over	_____

Answers to Odd Problems (For Self-Checking Purposes)

Chapter 1. Place Values

Place Values

1)	10th	7)	10,000s	13)	10th	19)	10,000s	25)	0
3)	100th	9)	100th	15)	100s	21)	2	27)	7
5)	100s	11)	10,000th	17)	10s	23)	3	29)	8

Comparing Numbers

1)	4.786	9)	49300	17)	26.3	25)	2.65	31)	8.100
3)	6.2	11)	15	19)	53	27)	7676	33)	1002.0
5)	700	13)	2173	21)	151.1	29)	8100	35)	.1010
7)	486	15)	6200	23)	2577				

Rounding Numbers

1)	57,000	1)	57,300	1)	57,260	1)	5	1)	4.8
3)	3,000	3)	3,200	3)	3,180	3)	52	3)	51.7
5)	2,000	5)	2,200	5)	2,180	5)	214	5)	213.6
7)	877,000	7)	876,500	7)	876,520	7)	4	7)	4.0
9)	27,648,000	9)	27,647,500	9)	27,647,530	9)	5	9)	5.2

1)	4.78	1)	4.777	1)	4.8	11)	3.755
3)	51.67	3)	51.673	3)	2.1	13)	150
5)	213.62	5)	213.617	5)	4	15)	360
7)	4.01	7)	4.006	7)	2100	17)	9.00
9)	5.15	9)	5.152	9)	21,800		

Reading and Writing Numbers as Words

17) 17 thousand 3 hundred 44 and 47 hundredths

19) 1 million 6 hundred 94 thousand 8 hundred 69

21) 37 and 97 hundredths

23) 8 hundred 82 and 22 hundredths

25) 23 and 6 thousand 3 hundred 75 ten-thousandths

27) 28 thousand 5 hundred 87

29) 16 and 33 thousand 6 hundred 85 hundred-thousandths

Chapter 2. Whole Number Operations

Addition

1)	31	7)	88	13)	5179	19)	90,373	23)	2760, 2895, 130, 5785, 560, 660, 1385, 170, 2775
3)	29	9)	54	15)	1689	21)	578, 28, 1394, 957		
5)	43	11)	5159	17)	203				

Subtraction

1)	6	7)	32	13)	3831	19)	320	23)	11, 3, 7, 16
3)	63	9)	27	15)	441	21)	1105, 19, 722, 506		
5)	68	11)	675	17)	586				

Multiplication

1) 70	7) 225	13) 6004532	19) 11,468,000	23) 196
3) 66	9) 192	15) 671,328	21) 320, 80, 210,	
5) 42	11) 16192	17) 5733	235, 1320	

Division

1) 6 R 36	7) 16 R 44	13) 51 R 7	19) 912 R 12
3) 122 R 4	9) 28 R 32	15) 83 R 39	21) 29
5) 6 R 8	11) 202	17) 92 R 49	23) 2382

Mixed Operations

1) 45 hours	3) 33 rooms	5) $14	7) $147	9) 44 hours

Chapter 3. Decimal Operations

Addition

1) 125.403	5) 244.3951	9) 119	13) 1476.348
3) 1	7) 223.242	11) 400.06	15) 4.0401

Subtraction

1) 26.769	5) 8961.7996	9) 1743.9	13) 7.9 mg, 6.132 mg, 3.4 mg
3) 1.016	7) .01	11) 1529.13	15) 1.70833

Multiplication

1) .006007	5) .040405	9) 89.7204	13) 34.978, 896.51, 23.40
3) 257.792	7) 8339.1	11) 1856.25	15) 42.5, 33.75, 40, 31.5

Division

1) 1.56	5) 8.04	9) 1.12	13) 10.79
3) 235.5	7) 4.18	11) 3.77	15) .23, .22, brand B

Mixed Operations

1) 356.28	5) $3.10, $3.16, Gladwells—$.06	9) $11.77
3) $.91/hr	7) 39.2 mi	

Chapter 4. Fraction Operations

Least Common Denominators

1) 35	5) 24	9) 60	13) $\frac{6}{9}, \frac{4}{9}$	17) $\frac{5}{6}$
3) 10	7) 40	11) $\frac{36}{84}, \frac{77}{84}$	15) $\frac{48}{60}, \frac{35}{60}$	19) $\frac{5}{6}$

Reducing Fractions

1) $\frac{1}{8}$	5) $\frac{1}{6}$	9) $\frac{2}{3}$	13) $\frac{5}{17}$	17) $\frac{2}{9}$
3) $\frac{1}{4}$	7) $\frac{2}{7}$	11) $\frac{2}{5}$	15) $\frac{3}{4}$	19) $\frac{1}{8}$

Mixed Numbers and Improper Fractions

1) $\frac{30}{7}$	9) $\frac{83}{8}$	17) $\frac{19}{4}$	25) $8\frac{2}{5}$	33) $13\frac{1}{3}$
3) $\frac{32}{6}$	11) $\frac{13}{3}$	19) $\frac{31}{7}$	27) $2\frac{2}{7}$	35) $9\frac{1}{6}$
5) $\frac{13}{4}$	13) $\frac{23}{4}$	21) $3\frac{3}{4}$	29) 2	37) $3\frac{3}{5}$
7) $\frac{20}{3}$	15) $\frac{25}{9}$	23) $4\frac{2}{5}$	31) $5\frac{1}{3}$	39) $22\frac{1}{2}$

Addition

1) $8\frac{14}{15}$	3) $8\frac{39}{70}$	5) $3\frac{17}{36}$	7) $1\frac{1}{20}$	9) $8\frac{17}{40}$

Subtraction

1) $\frac{1}{8}$	5) $\frac{23}{60}$	9) $\frac{5}{24}$	13) $3\frac{4}{9}$
3) $\frac{11}{70}$	7) $\frac{1}{5}$	11) $2\frac{1}{2}$	15) $2\frac{7}{10}$

Multiplication

1) $1\frac{7}{18}$ 5) $1\frac{1}{12}$ 9) $1\frac{3}{8}$ 13) $26\frac{2}{7}$ 17) $26\frac{19}{20}$
3) $\frac{21}{40}$ 7) $1\frac{9}{16}$ 11) $3\frac{39}{50}$ 15) $16\frac{1}{4}$ 19) 9

Division

1) $2\frac{1}{3}$ 5) $\frac{3}{11}$ 9) 1   13) 4
3) 2   7) $1\frac{31}{33}$ 11) 8   15) 11

Mixed Operations

1) 49 oz   5) $167\frac{13}{16}$ 9) $\frac{3}{5}$
3) $15\frac{1}{2}, 11\frac{3}{8}, 12\frac{9}{16}, 11\frac{1}{4}$ 7) $27\frac{5}{8}$

Chapter 5. Combined Operations

Percent to Decimal

1) .55   7) .49   13) .10   19) .00361   25) .45
3) .65   9) .74   15) .081   21) .71   27) .01
5) 1.00   11) 1.255   17) 3.722   23) .05   29) .036

Decimal to Percent

1) 58.7%   7) 11.1%   13) 530%   19) 88%   25) .1%
3) 73.14%   9) 60%   15) 722%   21) 6400%   27) .5%
5) 400%   11) .7%   17) 333%   23) .06%   29) 1350%

Decimal to Fraction

1) $4\frac{14}{25}$ 7) $\frac{2}{25}$ 13) $7\frac{1}{2}$ 19) $\frac{1}{25}$ 25) $\frac{22}{25}$
3) $25\frac{1}{10}$ 9) $\frac{3}{50}$ 15) $\frac{13}{50}$ 21) $\frac{7}{1000}$ 27) $\frac{11}{50}$
5) $\frac{77}{500}$ 11) $3\frac{7}{10}$ 17) $\frac{4}{25}$ 23) $\frac{21}{50}$ 29) $\frac{1}{40}$

Fraction to Decimal

1) .8   5) 4.3   9) 3.2   13) 1.5
3) .1   7) .7778   11) .875   15) .95

Percent to Fraction

1) $\frac{1}{2}$ 7) $\frac{2}{25}$ 13) $\frac{7}{50}$ 19) $\frac{107}{400}$ 25) $\frac{3}{25}$
3) $\frac{23}{200}$ 9) $1\frac{4}{25}$ 15) $\frac{13}{20}$ 21) $\frac{3}{50}$ 27) $\frac{1}{400}$
5) $\frac{1}{25}$ 11) $\frac{17}{40}$ 17) $\frac{13}{25}$ 23) $\frac{29}{400}$ 29) $\frac{19}{50}$

Fraction to Percent

1) 57.14%   5) 11.11%   9) 25%   13) 12.5%
3) 50%   7) 180%   11) 175%   15) 92.86%

Percent-Base

1) 9   5) 40%   9) 150   13) 25%   17) 47%
3) 8   7) 9.6   11) 35%   15) 252   19) 121

Mixed Operations

1) .75, 75%   5) $\frac{33}{200}$, .165   9) 3.12, $3\frac{3}{25}$ 13) 4.05, 405%   17) .3125, 31.25%
3) $\frac{21}{400}$, .0525   7) .06, 6%   11) $4\frac{1}{2}$, 450%   15) $\frac{1}{2500}$, .0004   19) $\frac{3}{500}$, .6%

Chapter 6. Ratios and Proportions

Defining a Ratio

1) $\frac{1}{3}$, .3333   5) $\frac{4}{12}$, .3333   9) $\frac{7}{10}$, .7   13) 77:100, etc.
3) $\frac{10}{25}$, .4   7) $\frac{15}{20}$, .75   11) 23:50, etc.   15) 27:20, etc.

Defining a Proportion

1)	T	5)	Not T	9)	T	13)	T	17)	T
3)	T	7)	Not T	11)	Not T	15)	T	19)	Not T

Solving Proportions

1)	$x = 15$	5)	$x = 8$	9)	$x = 2.45$	13)	$x = 4$	17)	$x = 1$
3)	$x = 9$	7)	$x = 20$	11)	$x = 2$	15)	$x = 15$	19)	$x = 12$

Chapter 7. Measurement

Volume Measurement

1)	12	5)	3	9)	2	13)	$\frac{1}{8}$	17)	6
3)	2	7)	$\frac{1}{2}$	11)	4	15)	16	19)	2

Length Measurement

1)	5280	5)	30	9)	12	13)	63360	17)	74"
3)	4	7)	$5\frac{1}{2}$	11)	$\frac{1}{2}$	15)	96	19)	24 yd

Mass Measurement

1)	48	5)	67.2	9)	62	13)	71	17)	83 oz
3)	38	7)	7.5	11)	16	15)	11	19)	9.5 lb

Chapter 8. Metric System

Prefixes and Abbreviations

1)	kg, dl, dkm, l, mm, cg, hl, mg, g, dkl	3)	kg, hl, dm, ml, g, dkm, cl, km, dkg, l

Volume Measurement

1)	2900	5)	14	9)	294.7	13)	.00026	17)	100
3)	374.23	7)	60	11)	9.326	15)	8116	19)	2000

Length Measurement

1)	18760	5)	29000	9)	43000	13)	7.676	17)	6.921
3)	4.25	7)	.00498	11)	1438.2	15)	11.19	19)	1.75

Mass Measurement

1)	8630	5)	436000000	9)	86300	13)	.5611	17)	4650
3)	4.25	7)	.4231	11)	.0091164	15)	534000	19)	4000

Metric/English Conversions

1)	3, 90, 360, 195, 1125, 15, 960, .667	7)	58.6 kg
3)	5.46, .5, .62, .15	9)	720 cc
5)	2.724, 68.1, .448, .112	11)	1.3 pt

Chapter 9. Roman Numerals

Converting Roman to Arabic

1)	747	7)	900	13)	18	19)	105	25)	2754
3)	94	9)	14	15)	420	21)	996	27)	1457
5)	26	11)	1950	17)	34	23)	89	29)	24

Converting Arabic to Roman (answers are given in shortest form)

1)	DXXII	7)	X	13)	CDLVII	19)	CCCLXII	25)	CLXIII
3)	MIC	9)	$\overline{\text{V}}$CCCLXXVIII	15)	LXXIX	21)	DIC	27)	CLV
5)	LXXXII	11)	MCDXV	17)	LXXVII	23)	DCCCXIII	29)	$\overline{\text{VII}}$DCCCXII

Chapter 10. Medication Dosages

Adult Oral Dosages

1) 10 ml 3) $\frac{1}{2}$ tab, 15 tab total 5) 2 tab 7) $\frac{1}{2}$ caplet 9) $\frac{1}{2}$, 2

Adult Parenteral Dosages

1) .02, .05, .05, .1 (all ml) and 21 weeks 5) 6.25 ml 9) 9 ml
3) 3.75 ml 7) 50 ml

Calculating IV Flow Rates

1) 3.125 gtt/min 3) .694 gtt/min 5) 46.8 gtt/min

Calculating Children's Dosages

1) Clark's; 160 mg 5) Rule E; 100, 25 mg
3) Young's; 47.83 mg 7) Clark's; 400, 200 mg

Chapter 11. Vital Signs

Temperature Reading

1) oral/rectal 7) 97^8 11) 97 15) 101^8 19) 100^4
3) $98°$ 9) 99^8 13) 99^2 17) 97^4 21) see illustration
5) electronic

Temperature Conversion

1) 134.06 F 7) 59 F 11) 37 C 15) 24.78 C 19) 53.6 F
3) 8 C 9) 41.11 C 13) 212 F 17) 100.4 F 21) 99.5 F, yes
5) 111.2 F

Recording Temperature/Pulse/Respiration

See teacher's key for this section

Reading a Sphygmomanometer

1) 100 7) 122 13) 60 19) 108
3) 108 9) 128 15) 74 21) 124
5) 112 11) 136 17) 90 23) 134

Recording Blood Pressure

See teacher's key for this section

Chapter 12. Intake and Output

See teacher's key for this section

Chapter 13. Money

Making Change

1) Say $29, give $1, say $30, give $10, say $40.
3) Say $6.95, give $.05, say $7, give $1, say $8, give $1, say $9, give $1, say $10.
5) Say $2.88, give $.01, say $2.89, give $.01, say $2.90, give $.10, say $3, give $1, say $4, give $1, say $5, give $5, say $10.
7) Say $86.75, give $.25, say $87, give 1, say $88, give $1, say $89, give $1, say $90, give $10, say $100.
9) Say $39, give $1, say $40.
11) $.10, $.10, $.25, $.25, $.25, $1, $1, $1
13) $.01, $.05, $.10, $.25, $.25, $.25, $1, $1
15) $.10, $.10, $1, $10, $20
17) $1, $10
19) $.05, $1, $5

Collecting Money

See teacher's key for this section

Balancing a Cash Drawer

	1)	Cash in Drawer	$402.81	3)	Cash in Drawer	$604.00	5)	Cash in Drawer	$441.54
		+ Cash Pd. Out	4.25		+ Cash Pd. Out	1.30		+ Cash Pd. Out	6.00
		Total Cash	407.06		Total Cash	605.30		Total Cash	447.54
		Less Change	40.00		Less Change	40.00		Less Change	35.00
		Cash rec'd, #1	367.06		Cash rec'd, #1	565.30		Cash rec'd, #1	412.54
		Cash rec'd, #2	376.06		Cash rec'd, #2	570.30		Cash rec'd, #2	302.04
		Amt. *short*/over	9.00		Amt. *short*/over	5.00		Amt. short/*over*	110.50

Chapter 14. Time

Elapsed Time

1)	5:57	5)	8:16	9)	6:27	13)	4:04, 4:00, 8:04	15)	3:47, 3:37, 7:24
3)	7:11	7)	5:19	11)	3:51, 3:56, 7:47				

Time Sheets

1) 8:45, 8:00, 6:30, 8:45, 9:15, 41:15   5) 4:45, 6:15, 7:30, 5:15, 8:30, 32:15

3) 8:30, 9:30, 8:45, 8:15, 8:15, 43:15

Chapter 15. Income

Salary

1)	219.84	5)	557.70	9)	587.52	13)	3041.33
3)	305.78	7)	352.00	11)	287.98	15)	183.75

Deductions

1)	16.06	5)	40.74	9)	39.24	13)	24.02, 24.75
3)	30.59	7)	27.27	11)	21.21	15)	45.99

State Income Tax

1)	47.50	5)	724.63	9)	395.40	13)	461.67, 19.24
3)	416.81	7)	416.00	11)	18,772, 191.58	15)	20,590.68, 780.62, 15.01

FICA Tax

1)	79.61	5)	31.47	9)	91.17	13)	16.44
3)	91.25	7)	158.99	11)	46.18	15)	50.92

Voluntary Deductions

1)	3.62, 2.48, 3.38, 3.53, 78.01	7)	36.95
3)	17.48, 10.82, 39.58, 46.04, 113.92	9)	4.04, 1.86, 2.86, 4.07, 77.83
5)	7.31, 3.69, 1.73, 22.73		

Net Pay

1) 291.84, 29.97, 11.82, 21.92, 63.71, 228.13

3) 228.00, 10.98, 6156.00, 9.12, 17.12, 40.02, 187.98

5) 210.00, 25.04, 11.03, 15.77, 51.84, 158.16

Earning Statement

1) FIT 29.97, FICA 21.92, State 11.82, Total Deductions 63.71, Regular Pay 243.20, Overtime Pay 48.63, Gross Pay 291.84, Net Pay 228.13

3) Regular Pay 228.00, Gross Pay 228.00, YTD (year to date) 6156.00, FIT 10.98, YTD 296.46, State 9.12, YTD 246.24, FICA 17.12, YTD 462.24, Other 2.80, YTD 75.60, Net Pay 187.98, YTD 5075.46

5) FIT 25.04, FICA 15.77, State 11.03, Total Deductions 51.84, Regular Pay 210.00, Gross Pay 210.00, Net Pay 158.16

Paychecks

1–5) See teacher's key for this section

7) FIT 23.14, FICA 26.51, State 9.71, Total Deductions 59.36, Regular Pay 312.00, Overtime Pay 40.95, Gross Pay 352.95, Net Pay 293.59

9) FIT 26.24, FICA 28.04, State 10.27, Total Deductions 64.55, Regular Pay 312.00, Overtime Pay 61.43, Gross Pay 373.43, Net Pay 308.88

Chapter 16. Office Skills

Numerical Filing

1) 167, 174, 176, 177, 189, 198, 1137, 1248, 1361, 1637, 1714, 1737, 1741, 1748, 1763, 1776, 1789, 1822, 1828, 1889, 1899, 1988, 11718, 12745, 13781, 16370, 16371, 17417, 17482, 18229, 18292, 19003

3) 1234564, 1236431, 1267537, 1345435, 1347123, 1347413, 1347431, 1347434, 1347871, 1347874, 1373483, 1374317, 1378976, 1384134, 1473134, 1713741, 2312341, 2345511, 2345654, 2345676, 2347245, 2347423, 3456765, 3457834

5) 1236–5, 12347–5, 12362–5, 12363–5, 12367–5, 12473–5, 12742–5, 13873–5, 17246–5, 18324–5, 1263–6, 12362–6, 12728–6, 12742–6, 12754–6, 16134–6, 16326–6, 17347–6, 17437–6, 17437–6, 18534–6

Time Management

See teacher's key for this section

Writing Receipts

See teacher's key for this section

Chapter 17. Personal Finance

Personal Checking

1) Final Balance $1663.52

Personal Savings

1) Deposit $179.42

3) Deposit $1905.82

5) fill in blanks

7) fill in blanks

9) Deposit $1789.17

Credit

See teacher's key for this section

Utility Bills

See teacher's key for this section

Purchasing a Car

1) 9494.64

3) 7379.41

5) 14,870.58

7) 1346.91

9) dp = 1696.24, loan = 5088.71

Monthly Budget

1) 1314.17, 420.53, 197.13, 197.13, 65.71, 65.71, 91.99, 65.71, 65.71, 78.85, 65.70

3) 2360.00, 755.20, 354.00, 354.00, 118.00, 118.00, 165.20, 118.00, 118.00, 141.60, 118.00

5) 5178.00, 1656.96, 776.70, 776.70, 258.90, 258.90, 362.46, 258.90, 258.90, 310.68, 258.90

7) 1950.33, 624.11, 292.55, 292.55, 97.52, 97.52, 136.52, 97.52, 97.52, 117.02, 97.52

9) 792.00, 253.44, 118.80, 118.80, 39.60, 39.60, 55.44, 39.60, 39.60, 47.52, 39.60

Index